The Ancestral Rocky Mountains of Utah

By William A. Szary

Book Cover: The Yampa Plateau of northeastern Utah is part of the Wyoming Shelf in the Uinta Basin. The upper plateau represents Mesozoic rocks deposited on top of Paleozoic rocks to form the plateau surface.

Library of Congress Catalog in Publications Data:

Szary, William A. The Ancestral Rocky Mountains of Utah

Includes references

ISBN 13: 9798559582863

Earth2Energy Educational Publishing
Port Richey FL 34668

Earth2Energy is a Registered Trademark

Table of Contents

Chapter 1. Tectonic and Geologic Setting

The boundary between the Colorado Plateau and Southern Rocky Mountain provinces lies somewhere within the western Colorado-eastern Utah region. It is arbitrary in the geological sense because the stratigraphy and tectonic relationships remain relatively constant throughout the area.

The area of this chapter lies astride the Uncompahgre Uplift segment of the Ancestral Rockies orogenic system. To the west of the Ancient Uncompahgre fault block lies the northwest trending Paradox basin and to the east is the strikingly similar Eagle evaporite basin, both of late Paleozoic age. Bounding the region on the east are the marginal Gore and Mosquito faults of the Front Range segment of the Ancestral Rockies. Within this vast area, the early Paleozoic stratigraphy is similar and essentially uniform within the paleo-tectonic basins suggesting that the two down warps were not so distinctly separated until after Mississippian time. However, in Pennsylvanian time the main uplift of fault blocks of the Ancestral Rockies abruptly separated and accentuated the Paradox and Eagle evaporite basins, although the stratigraphy of the two depressions remained remarkably similar. Both the Uncompahgre and Front Range uplifts became high topographic features and shed untold cubic kilometers of coarse arkosic sediments into the eastern reaches of the structural basins. Whether the two salt basins were interconnected in a common sea is still a matter of personal interpretation.

The margins of the Paradox basin usually are defined by the geographic extent of salt deposited during Middle Pennsylvanian time in the Paradox Formation. Consequently, there is little or no reflection of the buried basin at the surface, except for the salt diapirs in the "Paradox fold and fault belt." The basin is bounded on the northeast and east by the Uncompahgre Uplift and is surrounded elsewhere by paleo-tectonically controlled shallow-water shoals. The ovate basin has a northwesterly orientation, extending from Durango, Colorado and Farmington, New Mexico on the southeast to Green River, Utah on the northwest.

The Uncompahgre Uplift, which is best known for its exposures on the Uncompahgre Plateau and Black Canyon of the Gunnison River, extends for some 700 km northwestward from near Santa Fe, New Mexico almost to Provo, Utah. It is complexly fault bounded on its southwestern margin and tilts gently eastward into the Eagle basin (**Figure 1**).

Figure 1. Black Canyon of Gunnison Colorado represents the Uncompahgre Uplift is fault bounded on the southwestern side, tilted gently into the Eagle Basin. Precambrian metamorphics are overlain unconformably by sedimentary deposits of Mesozoic age.

Where exposed, the crest of the tilted fault block is Precambrian metamorphic and igneous rocks overlain by various Mesozoic formations. Although it has a complex growth history, the Uncompahgre Uplift was a dominant feature throughout most of Permo-Pennsylvanian time.

The Eagle evaporite basin lies generally between the tilted eastern margin of the Uncompahgre Uplift and the fault-bounded Front Range Uplift on the east. The generally northwest-trending structural depression is a diminutive copy of the Paradox basin. The Eagle Valley Evaporites in the central part of the basin include gypsum and halite interbedded with drab-colored clastic rocks that inter-finger eastward with shelf carbonates and clastics of the Minturn Formation and arkose of the Maroon Formation. As in the Paradox basin, salt diapirs are present at a reduced scale near Glenwood Springs and Eagle, Colorado (**Figure 2**).

Figure 2. The Roaring Fork River south of Glenwood Springs Colorado exposes gypsum and halite evaporites (white) between drab red colored clastics of the Eagle Basin.

Basement Framework

Although much has been said about the Laramide deformation of the Colorado Plateau and Southern Rocky Mountain provinces, the fact is that the structural fabric of the region was fixed by Late Precambrian time and repeated rejuvenations of the basement structure, including the Ancestral Rockies and Laramide episodes, only modified the original framework.

It was sometime around the summer of 1,700 m.y.b.p. that activity got underway on two major shear systems that transect the Paradox basin and Uncompahgre Uplift as we know them today. One, the dominant northwest-trending swarm of faults that passes through the San Juan Mountains of southwestern Colorado and on into the subsurface of the eastern Paradox basin may extend as far to the northwest as Vancouver Island, B.C. and toward the southeast into Oklahoma's Wichita aulacogen. The subordinate northeasterly swarm of faults forming the conjugate set of fractures extends from Grand Canyon, Arizona through the Colorado Mineral Belt to Lake Superior. The northwest-trending set, called the Olympic-Wichita lineament appears to have had right-lateral strike-slip displacement, movement took place between 1,720 m.y. and 1,460 m.y. in the San Juan Mountains. The northeast-trending Colorado lineament, dated at 1,700 m.y. displaced the basement rocks in a left-lateral sense. The two continental-scale shear systems form a conjugate set that intersect in the vicinity of Moab, Utah.

The basic fracture pattern outlined above could be formed when compressive forces were directed from the north (or south) in Precambrian time. Perhaps a more reasonable solution would be a right-lateral wrenching stress imposed on the western United States that was proposed in an "outrageous hypothesis." Whether the underlying mechanism is the "outrageous hypothesis" or north-south compression is irrelevant for this discussion. Recognition of the basic tectonic fabric provides a basis for understanding the "peculiar" structural features of the Colorado Plateau and Southern Rockies.

A review of the strain ellipsoid oriented to properly represent the conjugate fractures described above indicates that northerly-trending normal faults (east-west extension) should occur in conjunction with the wrench faults (**Figure 3**). It has been well established that the large monoclinal folds of the Colorado Plateau are drape structures over normal basement faults of Precambrian age. Thus the monoclines originated as third-order features of the stress field. Furthermore, the large bounding faults of the Colorado Rockies (the Gore and Mosquito faults along the Front Range Uplift and the Cotopaxi fault that bounds the Wet Mountains Uplift farther south) are north-northwest normal faults that originated in Precambrian time and appear to have resulted from the same basic mechanism. Early Paleozoic rejuvenation of these features expects to have easterly-trending folding and thrusting (north-south compression) which readily explains the "maverick" east-west trending thrusted anticlines of the Uinta Mountains. At least the very large Uinta Mountain arch had a Precambrian origin and fits the stress field described (**Figure 4**). Thus, the tectonic stage was set long before the Ancestral Rockies and Laramide orogenies, and most likely by about 1,700 m.y.b.p.

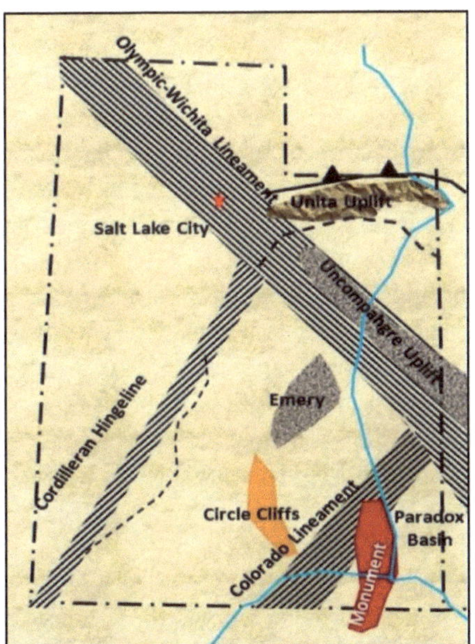

Figure 3. Location of the Colorado Plateau and relationship to major orthogonal set of basement lineaments. Northwesterly lineaments are right lateral, northeasterly lineaments are left lateral. Stress strain ellipsoid oriented such that maximum compressive stress is directed from the north.

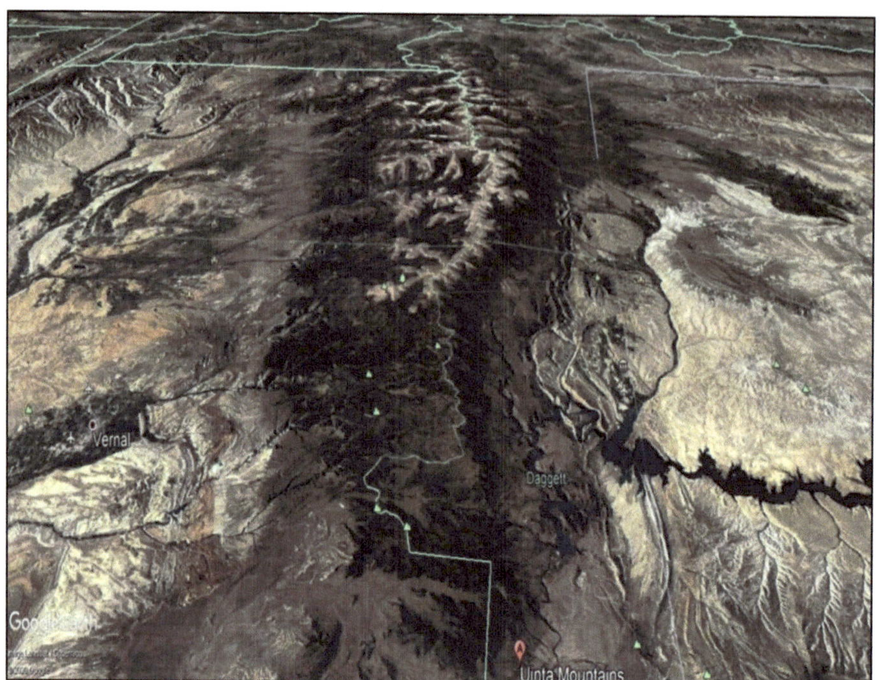

Figure 4. The Uinta Mountain arch can be traced from easterly tilted beds to the right along the flanks of the mountain core, and the westerly tilted beds to the left of the mountain on the flanks.

Fault Rejuvenation

The Colorado Plateau and Southern Rockies were relatively quiescent during early Paleozoic time. Evidence suggests that minor rejuvenations along the Olympic-Wichita lineament in the Paradox basin occurred during Cambrian, Devonian, and Mississippian times. Similar rejuvenations of tectonic activity along major structures in the vicinity of the Eagle basin and Front Range and Wet Mountains uplifts were described. Although early Paleozoic displacement on the faults was minor, sufficient vertical movement occurred to create local shoaling conditions and alter sedimentary facies on a local scale. This structural activity was responsible for isolating local reservoir facies in Late Devonian and Early Mississippian rocks of the central Paradox basin. Similar potential reservoir facies in Mississippian strata of the southern Eagle basin have been reported in close proximity to basement structures. It appears that repeated paleo-tectonic rejuvenation on a minor scale is favorable to the development of petroleum reservoirs prior to Pennsylvanian time.

Ancestral Rockies Uplift

The relatively mild nature of the early Paleozoic tectonic activity took a dramatic change by the beginning of Middle Pennsylvanian (Atokan) time. Vertical displacement along the basement faults of the Paradox basin and Southern Rockies began slowly, but picked up momentum as Middle Pennsylvania (Desmoinesian) time progressed.

The major uplifts of the Ancestral Rockies developed during this time as tilted fault blocks rising from Precambrian wrench faults and major normal faults became major source areas for coarse arkosic clastic sediments. The Uncompahgre, Front Range and Wet Mountains uplifts arose to considerable heights, shedding hundreds of cubic kilometers of coarse clastics to the adjacent basins. More than a thousand meters of arkosic debris accumulated along the eastern flanks of the Paradox and Eagle (Maroon) basins in Permo-Pennsylvanian time. Although vertical displacement definitely dominated the earlier wrenching stress field, the structural fabric of the basement remained essentially unchanged, and east-west extension of the crust was greatly enhanced. The increased tectonic activity resulted from the collision of the North American and South American-African plates at this time. However, the basic fabric remained essentially unchanged from the Precambrian pattern.

Uncompahgre-San Luis Uplift

The Uncompahgre Uplift had a complex history that is not completely understood because of poor exposures and limited subsurface data. Major uplift of the faulted highland began in Atokan time in a southern segment that extends from just east of Santa Fe, New Mexico northwestward to Ouray, Colorado. Relatively thick Atokan clastics in north-central New Mexico and southwestern Colorado attest to the early growth. This segment, the San Luis Uplift, continued its relatively modest rate of uplift well into Permian time, supplying clastic sediments to the Sangre de Cristo, Hermosa and Abo formations along the adjacent lowlands.

A pronounced westerly arc, or flex, occurs in the highland just east of Ouray, paralleling a similar westward flex in the adjacent Sneffels and Grenadier grabens of the San Juan Mountains. This arcuate bend in the basement faults is interpreted to be a left-lateral drag fold of enormous proportion caused by the Colorado lineament. It also appears to occur at the northwestern termination of the San Luis (Atokan) positive feature. From Ouray northwestward to the northwestern plunge of the surface Uncompahgre Plateau near Cisco, Utah, the middle (or Uncompahgre) segment of the uplift did not begin shedding massive amounts of arkose until Desmoinesian (Middle Pennsylvanian) time, but continued its positive tendencies well into Permian time (**Figure 5**). This northwesterly termination of the Uncompahgre Plateau also marks the emergence of the Colorado lineament into the Paradox basin. In other words, the Uncompahgre Uplift appears to have been segmented by the coupling of the Colorado and Olympic- Wichita lineaments.

Figure 5. The Uncompahgre uplift tilted northwesterly in the vicinity of Cisco Utah. Note the multiple folding events occurring in the left lower corner of the satellite imagery.

The northwestern element of the Uncompahgre Uplift, perhaps best termed the Book Cliffs segment extends from about Cisco to the over-thrust belt in the Wasatch Mountains east of Provo, Utah in the subsurface. Its existence is known from geophysical and sub-surface data underlying the southern flank of the Uinta Basin beneath the Book Cliffs. Thus, the rigid basement block provides a sill over which the structural basin sagged, creating the peculiar triangular shape of the basin. It did not shed significant amounts of arkosic debris until Early Permian time (**Figure 6**).

Front Range- Wet Mountains Uplifts

The complex growth history of the numerous individual segments of the Front Range-Wet Mountains uplifts of the Ancestral Rockies has been thoroughly documented and summarized. It will suffice here to reiterate that the history of the tectonic activity east of the Uncompahgre Uplift is similarly segmented and variably dated as that described for the Uncompahgre Uplift.

Figure 6. The Book Cliffs in Utah is the northwestern element of the Uncompahgre uplift. The southern flank of the Uinta Basin underlies the cliffs based on geophysical evidence.

Paradox Basin

The Paradox basin formed adjacent to the southwestern bounding faults of the Uncompahgre Uplift as a complimentary faulted depression. The deepest part of the basin lies immediately adjacent to the uplift, having stepped down structurally in a series of half-grabens from the western and southwestern shelves, or hinge line. Restricted marine circulation resulted in evaporite sedimentation throughout most of Desmoinesian (Middle Pennsylvanian) time. Salt deposition began in early Desmoinesian time in the deeper faulted troughs and slowly filled the basin, burying the basement faults by the close of the epoch. Simultaneously, the Uncompahgre Uplift was supplying ever increasing amounts of arkosic sediments to the northeastern margin of the basin. The highland was continuously, or episodically, rising throughout the remainder of Pennsylvanian time, as the basin sank along the continuously active basement faults. Most, if not all of the smaller structural features of the Paradox basin were in place and growing during Paradox time. These folds exist at the surface today, having been enhanced by Laramide compression, but they were sufficiently prominent in Middle Pennsylvanian time to significantly impede open circulation of marine waters and to control salt thickness. The generalized surface structure of the region is overlain with Paradox salt isoliths.

It is clear that the modern surface structures were actively developing around the periphery of the salt basin during the deposition of the salt and were actively affecting salt depositional patterns, as well as restricting marine circulation. These are all basement-related structures, and pre-date the Laramide orogeny.

Perhaps the most obvious and certainly the most interesting structural features of the Paradox basin are the large salt diapirs of east-central Utah and west-central Colorado. This broad region, termed the "Paradox fold and fault belt" overlies the deeper structural depression of the Paradox basin where depositional salt thickness may have reached 1500 to 2400 m. The oldest and thickest salt deposits lie in the basement-controlled half grabens described above, but massive salt flowage has totally obscured other depositional characteristics.

The salt anticlines are strongly elongated northwesterly, paralleling the Uncompahgre frontal faults and overlying deep-seated basement faults. As salt thickness reached about a thousand meters over the half-grabens, thick wedges of arkose were being deposited along the transition between the Uncompahgre highlands and the low-lying evaporite basin. The directional nature of the clastic overburden was a large factor in the initiation of salt flowage toward the southwest and the large basement faults, which are known to have vertical displacements of over 1800 m in Paradox Valley (**Figure 7**). Thus, the deep-seated faults were major buttresses to lateral salt movement. As the mobilized salt encountered the fault "scarps," flowage was directed upward and diapirism was initiated. Continued deposition of clastic overburden to the north-east caused salt to flow continuously or episodically until the supply was depleted in Late Jurassic time, as shown by repeated angular unconformities and pinch-outs along the flanks of the structures. The source of the ever-thickening clastic overburden on the Uncompahgre Uplift was buried by Late Triassic and Jurassic sedimentation as the supply of salt diminished, marking the close of the diapiric activity.

Eagle Evaporite Basin

The poorly known Eagle evaporite basin, lying between the Uncompahgre and Front Range uplifts, has a similar history of subsidence and sedimentation as the Paradox basin. A paucity of subsurface data and limited exposures of Middle Pennsylvanian deposits provide only limited data, and interpretations are therefore highly generalized.

As previously outlined, basement faulting with a history of repeated rejuvenation is similar, if not identical, to the Paradox basin. Coarse clastic deposits were derived from the Front Range Uplift, and perhaps others, to the east and southeast, and inter-fingered with marine clastics and carbonates of the Minturn Formation of the Central Colorado Trough and the Eagle Valley Evaporites in the northwestern Trough. As in the Paradox basin, algal bioherms of Middle Pennsylvanian age grew along the margins of the evaporite basin.

The extent and thickness of salt deposits in the restricted basin are only poorly known, and the nature of sub-salt structure is unknown due to the limited number of deep wells.

Figure 7. Paradox Valley in Colorado is a breached anticline displaying the directional nature of the clastic overburden controlling salt flowage toward the southwest accompanied by large basement faults known to have vertical displacements of over 1800 m in Paradox Valley.

Two northwesterly trends of Eagle evaporite exposures in the vicinity of Glenwood Springs and the Eagle-Gypsum area have the characteristics of salt diapirs similar to the Paradox structures. Drilling indicates that salt is present in the structures, but their size and growth histories have not been determined (**Figure 8**). These salt anticlines may be more similar to the Paradox structures than is generally realized, and further deep exploration of the pre-salt section is certainly warranted.

Ancestral Rockies Demise

The Ancestral Rockies had probably reached their zenith by Early Permian time. They underwent denudation for about the first half of the period, supplying great quantities of arkosic sediments to the lowlands during their demise. The Permian red beds from the uplifts finally buried many of the long-existing positive features of the region, such as the Zuni, Defiance, Kaibab, and Nacimiento uplifts of the Colorado Plateau. The Monument Upwarp remained sufficiently positive to cause major facies changes along its flanks and crests, while marine Permian deposits buried the Late Pennsylvanian erosional surface on the Emery uplift underlying the San Rafael Swell.

Fine-grained red beds and eolian sandstones dominated the Triassic and Jurassic systems, with little evidence of structural growth having played a significant role. By Late Cretaceous time, the intra-cratonic seaways shifted eastward, effectively burying the present day Colorado Plateau and Southern Rockies in marine muds and sands.

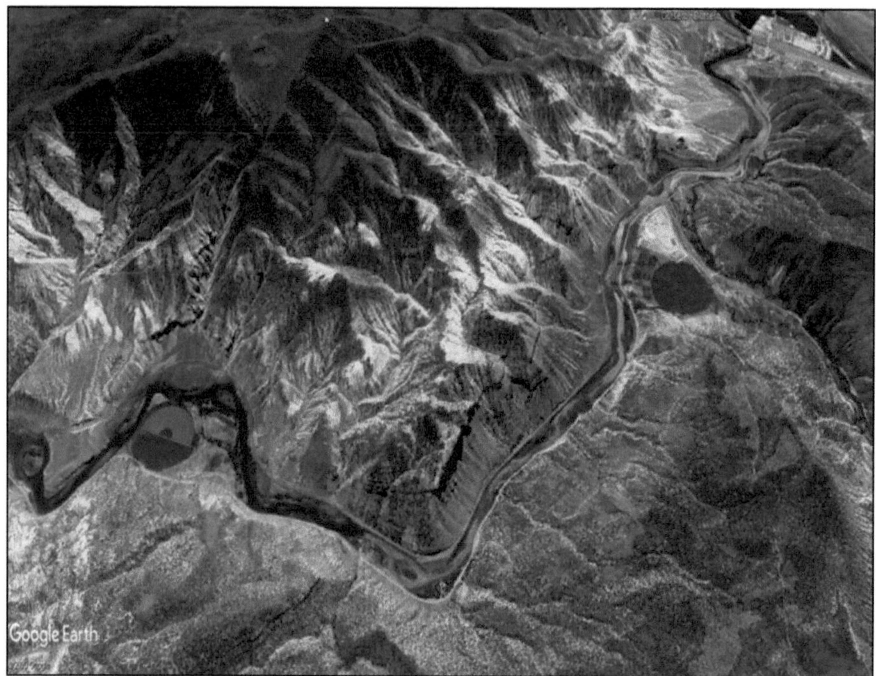

Figure 8. Dotero Colorado exposes gypsum evaporites in the hillsides west of Gypsum Colorado. Portions of the gypsum deposits are covered by clastics sediments shed into the Eagle Basin.

Chapter 2. The Uncompahgre Uplift in Utah

The origin of the Uncompahgre uplift is considered to be part of the Ancestral Rocky Mountain orogeny that started in Late Mississippian and continued through Middle Permian time. A series of basins and intra-cratonic block uplifts formed the Ancestral Rocky Mountains whose development has been attributed to two different tectonic models.

The first model relates the development of the Ancestral Rocky Mountains to the continent-continent collision of North America with South America–Africa producing the Ouachita-Marathon orogeny. This orogeny is considered the last suturing event to form the supercontinent of Pangea. As a result of this collision, preexisting zones of basement weaknesses were reactivated forming a series of basins and uplifts that define the Ancestral Rocky Mountains. The second tectonic model differs from the first one by proposing a northeast-southwest crustal shortening along the southeast margin (late Paleozoic Andean margin) of North America caused by northeast-dipping subduction. Evidence for this Andean margin comes from Central-Mexico which contained a volcanic arc during the Pennsylvanian and Permian indicating a northeast dipping subduction boundary to the south and west.

The Uncompahgre uplift is a mostly buried tectonic feature that separates the Paradox basin on the south from the Piceance basin on the north. The uplift covers an area of 7,000 km2 and trends northwesterly across the Utah-Colorado border at the Grand-Uinta County line in Utah. It is outlined on both the northeast and southwest by major fault zones called the Garmesa and Uncompahgre fault zones, respectively. On the north, the Uncompahgre uplift is bounded by the Piceance basin formed mainly in the Tertiary. The Piceance basin extends from a thrust belt in north central Utah on the west to the southern Park Range and Sawatch uplift in northwestern Colorado on the east. The northern boundary of the basin is defined roughly by the Uinta Mountains, and the southern boundary is located along a line north of the axis of the Uncompahgre uplift. Bounding the Uncompahgre uplift on the south, the Paradox basin is well known for the interesting salt structures that it contains. The deepest part of the basin lies immediately adjacent to the uplift, having stepped down structurally in a series of half grabens from the western and southwestern shelves.

Geological Background

The first recognizable tectonic activity in the Uncompahgre uplift region initiated in late Precambrian time with left-lateral wrench movement along the Uncompahgre fault zone to the south and the Garmesa fault zone to the north. This event was apparently followed by a period of relative quiescence, as no important tectonic activity is reflected in the early Paleozoic marine rocks around the periphery of the uplift.

In addition, the main phase of deformation in the Uncompahgre uplift took place during the Pennsylvanian and Permian, when approximately 3 km of coarse Hermosa and Cutler Formations arkose was dumped southward into the Paradox basin along the Uncompahgre fault zone. Pennsylvanian and Permian facies change from normal marine carbonates to non-marine red beds and clastics toward the central uplift along both its flanks, reflecting this tectonic uplift event.

Pre-Triassic Rocks: The Emery Uplift

Rocks of Pennsylvanian and Permian age are exposed below the Moenkopi Formation of Triassic age in parts of the San Rafael Swell, and oil tests show that sedimentary rocks of Pennsylvanian, Mississippian, Devonian, and Cambrian age lie between the base of the Pennsylvanian and the basement complex. The oldest rocks exposed probably belong to the Lower Permian Elephant Canyon Formation and possibly include the uppermost part of the Pennsylvanian Hermosa Formation. These rocks are exposed in Straight Wash in sec. 19, T. 23 S., R. 13 E., and consist from base to top about 100 feet of massive locally cherty finely crystalline limestone and dolomite and 260 feet of interbedded sandstone and carbonate rocks. They are overlain by more than 500 feet of flat and cross bedded sandstone that belongs to the Coconino Sandstone.

An interval about 50 feet thick below the Coconino consists of interbedded tan-weathering limestone and cross bedded sandstone and appears to be transitional between the Elephant Canyon (?) and the Coconino formations. Similar rocks form the base of the Coconino Sandstone in the Black Box (canyon) of the San Rafael River near the crest of the San Rafael anticline (**Figures 9 & 10**).

The Coconino Sandstone is mainly large-scale cross bedded fine grained sandstone and subordinately flat-bedded sandstone, particularly near the top of the formation. It is about 650 feet thick in the Black Box of the San Rafael River and appears to be about 880 feet thick in the Blackwood and Nichols 1-28 test well, sec. 28, T. 24 S., R. 10 E. (American Stratigraphic Co. log 705). It is overlain unconformably by the Kaibab Limestone which consists mainly of limestone, but contains thin shaly and sandstone beds near its base. The Kaibab ranges from 0 to about 85 feet in thickness in the swell. The correlations of the Coconino Sandstone and Kaibab Limestone of the San Rafael Swell with the type sections of these formations in the Grand Canyon area have long been regarded as tenuous. The sandstone-limestone sequence was similar not only to the Coconino and Kaibab in the southern Colorado Plateau but also to the Weber and Park City succession of north-central Utah. The Kaibab of the San Rafael Swell is younger faunally than the type Kaibab and that the type Coconino thins progressively northward, probably pinches out and is not north of the Utah-Arizona border.

Figure 9. The southeastern section of the San Rafael swell exposes the Elephant Canyon Formation. View is facing north along Straight Creek, Utah.

Figure 10. Lower Box Canyon exposes portions of the Elephant Canyon Formation in the lower cliffs overlain by Triassic red beds in the upper section.

The Kaibab of the San Rafael Swell rested unconformably on the Cedar Mesa (?) Sandstone Member of the Cutler Formation. Stratigraphic relations in the large number of oil tests recently drilled in the Green River desert east of the swell indicate that the Coconino of Utah is at least partly correlative with the White Rim Sandstone Member of the Cutler Formation. The Kaibab of Utah is probably in part correlative with the Park City Formation (**Figure 11**).

Chapter 3. Ancestral Rocky Mountains Load Induced Subsidence

Introduction

The Pennsylvanian–Permian Ancestral Rocky Mountains (ARM) of the west-central U.S. formed a collection of largely crystalline basement-cored highlands that shed debris into adjacent basins in western equatorial Pangea far from any recognized plate boundary. The term "Ancestral Rockies" arose nearly a century ago in recognition of the thick, coarse grained strata that wedge toward Precambrian basement regions of the modern Rockies. Many of these paleo-highlands are bounded by high-angle Pennsylvanian-aged faults reflecting significant (several kilometer) dip-slip offset, as well as lateral displacements. The core ARM uplifts are characterized by large structural displacements and thick (≥2 km), proximally conglomeratic mantles, and extend beyond the immediate Rocky Mountains region into Oklahoma. Here, ARM structures coincide spatially with much older structures linked to the Precambrian–Cambrian rifting of the Rodinian supercontinent.

Figure 11. The Green River Desert exposes members of the Cutler Formation overlain by Triassic formations.

The ARM forms a classic example of intra-plate orogeny and remains enigmatic, although several authors have linked the orogenesis to far-field effects of the Marathon-Ouachita convergent margin. New data and reanalysis of existing data indicate that even the termination of the ARM orogeny is enigmatic.

It has been long accepted that the ARM highlands continued to rise from Middle Pennsylvanian through at least Early Permian time, and that subsequent erosional beveling associated with isostatic adjustment over tens of millions of years ultimately obliterated the mountains by Triassic–Jurassic time. New observations of significant preserved paleo-relief on top of ARM uplifts challenge this view. This paleo-relief preservation is remarkable because it archives landscapes of great antiquity, and appears to record subsidence of highland and adjacent regions not previously recognized.

Geologic mapping, stratigraphic, petrologic, structural, and geophysical data are combined from some of the largest-magnitude ARM highlands and intervening regions documenting an episode of widespread subsidence that followed the tectonic apogee of the ARM orogeny. These observations are interpreted with documentation of a high-density crustal load underpinning the core ARM, modeling the possible effects of this load in light of the changing stress fields associated with ARM orogenesis. This analysis indicates that tectonic inheritance such as ancient mass loads in the crust or lithosphere should be considered as a previously unrecognized means to hasten the demise of orogenic highlands.

Geologic Setting: Early and Late Paleozoic Geologic Events of the North American Midcontinent

Major tectonic events that affected the North American midcontinent through Phanerozoic time include early Paleozoic (Early Cambrian) rifting associated with the breakup of the Rodinian supercontinent, and late Paleozoic (Pennsylvanian–Permian) compression associated with the assembly of the Pangean supercontinent. Although the latter event forms the focus of this section, the early Paleozoic events imparted a tectonic fabric that possibly influenced later deformation.

Early Paleozoic Southern Oklahoma Aulacogen

A series of igneous rocks and associated crustal-scale structures extending at least through Oklahoma and the Texas panhandle mark the trend of the so-called Southern Oklahoma Aulacogen (SOA). The SOA is a classic example of an intra-continental failed rift that was later tectonically inverted. Geologic studies indicate that a combination of thrusting and lateral movements occurred during its formation. Knowledge of the Cambrian extension is based on many studies of the bimodal igneous rocks exposed in the Wichita uplift of Oklahoma, regional relationships, and post-rift subsidence. Cambrian igneous activity resulted in intrusion of a voluminous gabbroic complex (Glen Mountains Layered Complex) and associated shallow intrusives (**Figure 12**).

Figure 12. Glen Mountains in Oklahoma exposes Cambrian intrusive activity of gabbroic and shallow intrusive complexes.

As much as 40,000 km3 of metaluminous silicic magmas were generated ca. 530 Ma, producing the Carlton Rhyolite Group (and intrusive Wichita Granite Group). This assemblage of Lower Cambrian granite, rhyolite, and gabbro forms the basement of southwestern Oklahoma and the Texas panhandle.

The SOA extended to the Uncompahgre uplift region of Colorado on the basis of distributed, but limited, Cambrian mafic intrusives. Recent geophysical studies corroborate this inference. Extensive petroleum exploration of the southern Oklahoma region provides good constraints on the post-rift thermal subsidence history which includes ~3 km of predominantly Ordovician carbonate strata preserved in uplifted blocks and within the axis of the proto-Anadarko basin. Following thermal subsidence and associated Ordovician sedimentation in the wake of Cambrian rifting, subsidence within the SOA region and greater interior North America slowed considerably. In southern Oklahoma, a relatively thin Silurian–lower Mississippian carbonate and shale section records this interval of tectonic quiescence. This period was followed by a Mississippian–Pennsylvanian subsidence event heralding the beginning of the present Anadarko basin and accompanying the Ouachita orogeny.

By latest Mississippian and Pennsylvanian time, the Ancestral Rocky Mountains orogeny commenced, recorded by uplift of various highlands and major subsidence and sediment accumulation within highland-adjacent basins. The core ARM uplifts exhibiting the largest-magnitude fault displacements across basin-bounding faults, and thickest mantles of locally derived coarse-grained conglomerate occur in Colorado and Oklahoma, most notably in the Wichita-Anadarko and Uncompahgre-Paradox systems. Within southern Oklahoma, the Wichita uplift–Anadarko basin system formed as a result of late Mississippian–Pennsylvanian ARM compression that inverted the failed Cambrian rift. The inversion structures are unusually large with at least 12 km of vertical separation between the Cambrian basement exposed in the Wichita Mountains and that are present in the subsurface of the adjacent Anadarko basin. This 12 km displacement results from the dual-phase history of the SOA Anadarko system wherein the basin contains 4–5 km of rift-related lower Paleozoic section, and an additional 7 km attributable to flexural induced subsidence related to Mississippian–Pennsylvanian compressional deformation during the ARM orogeny. The length of the SOA (~1500 km) approximates that of the Main Ethiopia and Kenya rifts combined, and the signature transects the prevailing northeast-southwest grain of the Meso-proterozoic basement of North America. Pennsylvanian strata are not well exposed in this region, but intense petroleum exploration has revealed the presence of thick uplift adjacent conglomeratic units. For example, the so-called Granite Wash within the subsurface along the Frontal fault zone (i.e. basin ward of the Wichita uplift) of Oklahoma and the Texas panhandle reaches thicknesses of 2–3 km and exhibits a well-known reverse stratigraphy representing the active unroofing of the Wichita uplift during Pennsylvanian time.

Minimal post-Paleozoic deformation in the craton region of the southern mid-continent has enabled nearly pristine preservation of this system. Therefore, the Anadarko basin area archives a complete record of early Paleozoic extensional to late Paleozoic compressional deformation. Similar relationships exist in Colorado. ARM tectonism resulted in as much as 8 km of vertical displacement documented from drilling and seismic data bordering the Uncompahgre uplift. Adjacent basins such as the Paradox basin accumulated several kilometers of syn-tectonic carbonate-clastic strata, and thick conglomeratic aprons mantling several uplifts.

Judging by sedimentation rates and structural biostratigraphic data, deformation that produced the Ancestral Rocky Mountains peaked in middle Pennsylvanian to earliest Permian time, albeit with some spatial variation such as Permian ages of deformation in the Marathon region of southwest Texas. Owing in part to this timing which coincides with that of the Ouachita-Marathon orogeny, the intra-plate deformation of the ARM was linked to far field effects of the Ouachita-Marathon orogeny, a model that seems particularly applicable for the Wichita-Anadarko system of Oklahoma.

In this model, the ARM intra-cratonic deformation stems from activation of preexisting weaknesses by propagation of far field stresses associated either with (south-dipping) subduction of promontories or wrenching of Laurentia as eastern parts of the Pangean suture locked.

The role of preexisting weaknesses suggests that faults associated with the ARM orogeny were formed initially by crustal rupturing during Proterozoic rifting. Others suggested a connection to events (convergence or mega-shear activity) along southwestern or western Laurentia, most notably for features in the far western ARM system.

Permian History of the Core ARM: Post-orogenic subsidence?

Permian on lap of the Wichita Uplift, Oklahoma

Lower Cambrian magmatic rocks in the Wichita Mountains of southwest Oklahoma form the largest surface exposure of the SOA. These units, uplifted and eroded during ARM tectonism now protrude through a mantle of Lower Permian red beds that otherwise extend across the region. Relative to the Rocky Mountain region in general, Oklahoma has remained less disturbed by post-Paleozoic tectonism. As demonstrated by apatite fission-track and auxiliary data from the Wichita Mountains and neighboring Anadarko basin, the region records ~800 m to 1.5 km (1–3 km inferred) of Permian–Jurassic burial before denudation began in the Late Jurassic, and ≤1.5 km of denudation in the late Mesozoic–Paleogene in response to tectonic and/or climatic influences, but no reactivation of old uplifts as occurred in the Rocky Mountains. Geologic relationships in the Wichita Mountains demonstrate that Permian strata on lapped paleo-relief on the Cambrian basement.

At the surface, stream drainages are visible today, carved into both Cambrian igneous basement and Cambrian–Ordovician carbonates, but these drainages commonly do not propagate head-ward into the superjacent Permian strata despite the less competent nature of the latter. Rather, drainages appear to have been beheaded (crosscut) by horizontal Lower Permian strata. Numerous shallow wells drilled in the 1950s reveal a carapace of Permian strata as much as 1 km thick on lapping the basement of the Wichita highland, thus revealing the magnitude of the on lap.

Seismic data corroborate and expand upon these surface and well-bore observations. The Mountain View thrust fault, the main fault of the Frontal fault zone, marks the boundary between the uplift and adjacent basin. This fault zone is well imaged seismically, and demonstrates profound displacement in the pre-Permian section, and an on lap by minimally deformed Early Permian strata that extend from the basin onto the uplift. These relationships have long been recognized and cited as evidence for a Pennsylvanian age for the deformation. However, the magnitude of the on lap (~1 km) and the extension of the on lap from the basin up onto and across the uplift have not been highlighted.

The subsidence history of the proximal Anadarko basin shows that subsidence associated with the late Paleozoic basin history continued into Permian time. This history depicts a rapid late Mississippian through late Pennsylvanian subsidence event, a less rapid but significant Early Permian event, and the eventual cessation of subsidence by middle Permian time. This subsidence analysis employs a composite stratigraphic section from wells in the fore deep of the basin, and includes a rare fore deep well with a complete log through the Permian section, thus capturing a subsidence event that is equivalent in age to the on lapping Permian strata previously documented. The surface observations of on lap of the Lower Permian strata onto Cambrian basement demonstrate that the Wichita Mountains as observed today represent a late Paleozoic (Early Permian) landscape undergoing exhumation for the first time since the Early Permian. That is, the paleo-mountains are being progressively exposed as erosion removes the friable Lower Permian mudstone units that mantle the paleo-landscape. The granitic hills of the Wichita Mountains display minimal evidence for modern erosion. For example, alluvial fan mantles of granitic material do not occur. Rather, the cross cutting relationships of Lower Permian strata across drainages carved into Cambrian basement date the landscape to the pre–Early Permian. This profound nonconformity has long been recognized, but its tectonic significance has largely escaped notice. Taken together with the subsurface data, these relationships document Early Permian subsidence that (1) abruptly postdates compressional uplift, and (2) extends beyond the fore deep of the Anadarko basin and onto the Wichita uplift. That is, the core of the uplift subsided along with the flanking basinal regions.

Permian On lap in the Uncompahgre Uplift, Colorado

In contrast to the minimally disturbed record of Permian burial in Oklahoma, the ARM paleo-uplifts of Colorado exhibit a complex history affected by Mesozoic burial, reactivation of uplifts during Cretaceous–Paleogene (Laramide) tectonism, and significant Neogene exhumation associated with landscape evolution and climate change in the Cenozoic Rocky Mountains. Of the Colorado ARM uplifts, however, the Uncompahgre system of western Colorado is comparatively well preserved with a carapace of relatively undeformed Mesozoic strata. This perhaps reflects its location within the larger Colorado Plateau which was relatively undisturbed by Laramide tectonism. The Ancestral Uncompahgre uplift was a large northwest-southeast trending feature that formed during the ARM orogeny and was separated from the adjacent Paradox basin to the southwest by a seismically imaged subsurface reverse fault system that exhibits as much as 8 km of up-to-the northeast vertical displacement. The modern Uncompahgre Plateau composes only a part of the ancient highland and consists of a Precambrian basement core mantled by Mesozoic strata, except where breached by Unaweep Canyon, a large gorge in the southwestern plateau that exposes the Precambrian crystalline core of the plateau. Clastic detritus eroded from the highland during Carboniferous–Permian tectonism accumulated in the adjacent Paradox basin to form the Cutler Formation.

The contact between the Cutler Formation and Precambrian basement along the southwestern front of the modern Uncompahgre Plateau is well exposed near Gateway, Colorado (**Figure 13**). In this location, the contact was a depositional on lap based on analysis of seismic data to the northwest where a zone of reverse faults in the subsurface was revealed. Recent detailed mapping confirmed this depiction of a Permian (Cutler) on lap contact, and expands the recognized extent of the on lap onto Precambrian basement. Furthermore, these mapping results indicate little deformation during the time recorded by the Cutler Formation now exposed at the surface, a point also emphasized. These relationships indicate that motion on the subsurface Uncompahgre fault largely ceased before deposition of the youngest (Permian) Cutler strata as suggested. At this location, the post-tectonic Permian Cutler Formation buries ~520 m of paleo-relief on Precambrian basement, observable in outcrop as documented on published maps. Furthermore, the Cutler Formation here projects into Unaweep Canyon, a hypothesized exhumed landscape with remnant Pennsylvanian–Permian fill. The age of the canyon fill is inferred from the combined evidence for its exclusively Precambrian provenance, Pennsylvanian–Permian palynomorph content, and shallow (late Paleozoic) paleo-magnetic inclinations.

Figure 13. The contact between the Cutler Formation and Precambrian basement along the southwestern front of the modern Uncompahgre Plateau is well exposed near Gateway, Colorado.

A pre-Mesozoic age for the Precambrian (inner) gorge of Unaweep Canyon is also inferred from geomorphologic relationships wherein Mesozoic strata crosscut tributary valleys carved in Precambrian basement, analogous to the cross cutting relationships noted for the Wichita Mountains.

Acceptance of the antiquity of Unaweep Canyon implies the preservation of at least 970 m of paleo-relief, as measured between the Precambrian– Mesozoic nonconformity contact on top of the Uncompahgre Plateau and the nonconformity contact between the inferred Pennsylvanian–Permian canyon fill and Precambrian basement encountered in a core hole in the base of the canyon. The on lap of the uppermost (exposed) Permian Cutler Formation onto Precambrian basement of the Ancestral Uncompahgre uplift records Permian burial of the Uncompahgre highland. This conclusion was first reached noting "After the [Uncompahgre] highland attained its maximum height and while the Cutler was being deposited, the highland began sinking—at least along its southwest flank". The inferred Pennsylvanian–Permian age of Unaweep Canyon is consistent with this conclusion, and increases the recognized magnitude of the on lap, from ≥500 m to nearly 1000 m (**Figure 14**).

Figure 14. Unaweep Canyon, Colorado cuts through the Uncompahgre Plateau at right angles exposing the unconformable surface on top of Precambrian basement and overlying clastics of Mesozoic age.

Passive post-tectonic erosional beveling of the highland was once thought to have produced the depositional on lap along the margin of the uplift. However, the burial extends on top of the paleo-uplift, well beyond the flanking regions indicating that the highland must have been buried by at least 970 m of sediment to preserve the observed paleo-relief. Accumulation of this stratal thickness on top of the highland and its paleo-relief in the absence of subsidence is difficult to conceive for the active margin of a compressional orogen.

The observations suggest that the highland subsided, and the on lapping Permian strata record burial of the uplift and of the fault along which the highland had been uplifted. That is, the uplift and surrounding regions underwent subsidence together. The thickness of the proximally exposed Cutler Formation, 965 m as indicated by the sequential measured sections of the Cutler Formation in its most proximal location against the uplift, provides a minimum amount of subsidence on the Uncompahgre highland, and closely approximates the 970 m of preserved paleo-relief within Unaweep Canyon. The inference of subsidence of the Uncompahgre uplift is also consistent with the observation that well data near the Gateway area show > 2000 m of Cutler strata on top of Precambrian basement, and thus subsidence of the up thrown block.

These data indicate that the paleo canyon was backfilled by the end of deposition of the uppermost Cutler Formation in Early Permian time. The Triassic Moenkopi Formation between the Permian Cutler and Triassic Chinle units in the proximal Paradox basin thins to almost nothing toward the highland with a slight angular (<5°) unconformity between the Cutler and Moenkopi units. These relationships could reflect gradual erosional beveling of a highland that persisted to Triassic time. Any such beveling should have produced a coarse clastic apron, yet no known strata exist that record this later (post-Cutler) history. Rather, the coarse-grained, locally derived conglomeratic aprons persist only into earliest Permian time. Alternatively, these relationships could reflect reduction (via subsidence) of the highland to a low-elevation surface against which the Moenkopi Formation on lapped and the ultimate disappearance of the uplift as a significant eroding source by middle Permian time. Taken together, the geologic relationships are most consistent with the interpretation that the proximal Cutler Formation records cessation of ARM uplift in the region followed by ~1 km of subsidence of the greater region, encompassing both the proximal Paradox basin and the adjacent Uncompahgre highland.

Permian on Lap and Provenance Relationships in the Greater ARM Region

Structural relationships analogous to those documented for the Uncompahgre and Wichita systems also exist in the intervening ARM regions. In southern Colorado, Pennsylvanian–Permian strata bury ARM faults and adjacent basement by 800–1000 m. Farther to the north, 315 m of the (Lower Permian) upper Fountain Formation record sedimentologic, structural and stratigraphic relationships that indicate that these strata postdate movement of the local basin-bounding fault, thus requiring a regional mechanism for accommodation. Post-orogenic burial of the piedmont of an orogen (e.g. Tertiary strata of the modern Front Range) can simply reflect strong erosion in the hinterland. However, preservation of substantial paleo-relief in the hinterland, rather than piedmont of both the Wichita and Uncompahgre systems precludes an explanation linked to denudation of the highland. Rather, the relationships documented here demonstrate Permian burial extending across the ARM highlands recording ~1 km of accommodation space (conservatively ignoring compaction effects). This implies that uplift of the mountains in this compressional orogen ceased, and the core highlands even beyond faulted flanking regions underwent subsidence beginning in Early Permian time.

Provenance data for Pennsylvanian–Permian strata flanking many ARM uplifts shed additional light on the timing of active uplift and erosion. For example, the well-documented Pennsylvanian–Lower Permian conglomeratic strata that mantle many of the core ARM uplifts of Colorado and Oklahoma are the hallmarks of ARM tectonism, cited for nearly a century as the basis for recognition of the ARM system. However, these conglomeratic units which record local highland erosion persist only into the Lower Permian section. Their subsequent disappearance aligns with detrital zircon provenance results from sandstone and siltstone of the greater region.

If the ARM uplifts persisted as significant sediment sources into Triassic–Jurassic time, then Mesozoic strata in ARM-proximal regions should exhibit a significant provenance signature reflecting an ARM source, but they do not. Rather, these Mesozoic strata reflect a dominant source from eastern Laurentia (e.g., Grenville basement exposed in the Appalachian-Ouachita system), and lack a significant signature from the crystalline basement (either the Yavapai-Mazatzal or Wichita provinces) coring the ARM uplifts, excepting a minor and inferred recycled population.

This provenance shift in Paleozoic strata of the Grand Canyon exhibits an evolution from a significant ARM source in the Pennsylvanian–earliest Permian to an Appalachian-Ouachita source in the later Permian. This shift is also captured in Lower Permian siltstone units of New Mexico and Oklahoma. The age and provenance of the continental strata on lapping the ARM faults and burying the hinterland paleo-relief provide approximate constraints on the timing of subsidence. Faulting ceased in the late Pennsylvanian and post-tectonic on lap began by Early Permian time. Furthermore, structural restoration of the Paradox basin fill indicates that for the Uncompahgre front, faulting had largely ceased and on lap initiated in earliest Permian time (between 293 and 284 Ma). More accurate constraints on timing await higher resolution dating of these continental units.

Geophysical Evidence for a Crustal Load Beneath the core ARM

Gravity data provide a clear picture of the extent of the SOA and the impressive mafic magmatism associated with the Cambrian rifting. In southwestern Oklahoma, both COCORP seismic reflection profiles and a large refraction, wide-angle reflection experiment were integrated with geologic, drilling, gravity, and magnetic data to produce a crustal model. Notably, the mafic complex exposed in the Wichita highland and associated with a >100 mGal gravity high occurs almost entirely within the upper plate of a large thrust zone manifested near the surface as the Mountain View fault. Regional gravity anomalies show that this mafic mass extends northwestward with slight (50 km) offset for ~500 km well into northeastern New Mexico before being disrupted by features associated with the Rio Grande Rift. Farther northwest, the San Juan volcanic field dominates the gravity field, but evidence for Cambrian rifting extending to the Uncompahgre highland, and geophysical studies of this highland indicate that it also is underlain by a high-density, high velocity body.

The amplitude of the anomaly across the Uncompahgre highland is ~50 mGal, but flanking strata are of lower density than those flanking the Wichita highland reduce the density anomaly needed within the Uncompahgre highland uppermost crust. Moreover, new field studies along this trend in the Wet Mountains of southern Colorado reveal a 40 mGal gravity high associated with Cambrian mafic and ultramafic complexes. These results, when merged with regional gravity and magnetic data, indicate that a large (>1000 km2) part of the Wet Mountains is cored by Cambrian mafic igneous rocks, an inference consistent with seismic refraction data.

It is thus inferred that the northwest trending gravity anomalies that begin in northeastern Texas are most prominent in southern Oklahoma, and extend to the Uncompahgre highland of Colorado, represent a high-density upper crustal load, the total length of which is ~1500 km. The potential effects of this load have not been previously addressed.

Modeling Effects of a Crustal Load

Today the Wichita highland is supported by a strong lithosphere that continues to support the load associated with a ~120 mGal Bouguer gravity anomaly high. Concentrating on geophysical manifestations of the SOA in southern Oklahoma (to avoid the Laramide complications of the Uncompahgre highlands), the density structure derived from gravity observations and seismic velocity interpretations along the profile highlighted were vertically integrated. The lithospheric load and thus the flexure forced by the Cambrian mafic and ultramafic bodies underneath were quantified (**Figure 15**).

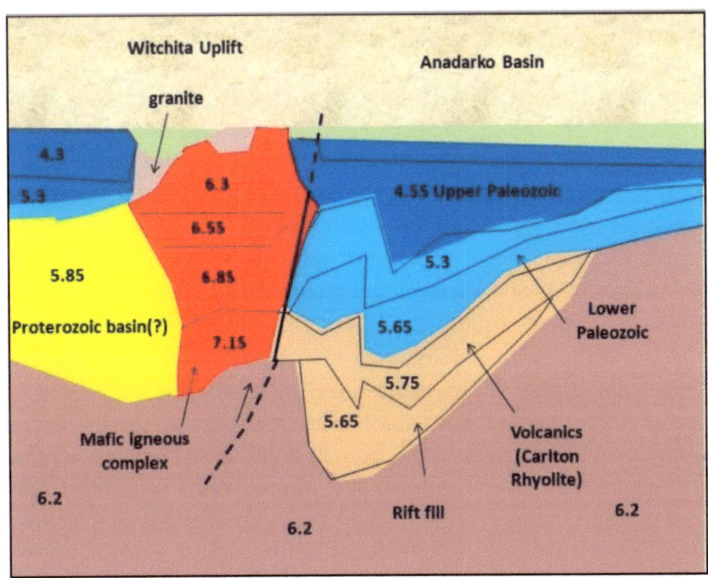

Figure 15. Velocity and density model of central portion of the SOA in Oklahoma and Texas. Numbers refer to seismic P wave velocities. The thrust fault line in the center represents the Mountain View Fault.

The disappearance of highland elevation in the core ARM, as recorded by preservation of Permian landscapes on top of the paleo-highlands records cessation of uplift followed by significant subsidence over a broad region. Mechanisms possibly capable of inducing such vertical motion include orogenic collapse, negative dynamic topography, or the influence of horizontal stresses. Any mechanism invoked to explain such motion must account for the significant vertical and areal extent of the subsidence, and the geologically abrupt onset following the apogee of ARM tectonism.

Orogenic collapse is well recognized as an important process in the evolution of mountain belts. Orogenic collapse, however, implies a large orogen with thick crust, usually accompanied by partial melting to weaken the thick crust, and typically postdates thickening by many millions of years. Such collapse involves gravity-driven flow that counteracts crustal thickening, reducing lateral contrasts in gravitational potential energy, and is commonly associated with extensional structures in the thickened crust and shortening in the foreland. These attributes do not occur in the ARM system, calling into question any role of orogenic collapse as traditionally defined. Dynamic topography refers to elevation differences caused by mantle flow, and has been invoked to explain large-scale continental flooding and exposure. Uplift and subsidence of large continental areas resulting from dynamic topography relate to mantle flow linked to the initiation and cessation of subduction. The potential appeal of dynamic topography for the ARM system derives from the presumed importance of the Ouachita-Marathon subduction system in ARM orogenesis; however, the system records southward-dipping subduction such that any potential influence of dynamic topography should have affected the Gondwanan rather than the Laurentian plate.

In-plane stress acting on an inhomogeneous crust provides an additional mechanism to explain large areas of vertical motion of the crust, including modulation of sedimentary basin formation. The mechanism of horizontal stresses acting upon a markedly inhomogeneous crust is most consistent with both the data and modeling results from the ARM system. The mafic keel of the SOA ARM system originated during the Cambrian rifting that produced the SOA, and underpins the core uplifts of the ARM discussed here. As reflected in the subsidence history of the earlier Oklahoma basin, and the later-evolved Anadarko basin, the loads and cooling caused subsidence of the SOA region into Mississippian time, but these loads subsequently acted as foci for ARM uplifts. Geodynamic modeling results are consistent with the hypothesis that relaxation, or lessening, of the compressional stresses that accompanied ARM orogenesis resulted in the cessation of uplift of the mountains, and active subsidence of the greater ARM region in response to the changing horizontal stress field in the context of the preexisting crustal in-homogeneities.

This mechanism provides a means to hasten the removal of highlands by creating positive accommodation in the hinterland, thus leading to long-term preservation of ancient highlands otherwise destined for erosional eradication. The core ARM highlands did not succumb to isostatically induced erosional beveling that reduced relief over tens of millions of years, for example as in the Appalachian orogen, because this was not a plate-margin collision associated with major crustal thickening. Therefore, once the stresses that induced ARM orogenesis ceased, the effects of the underlying mass load acted to reduce elevation through subsidence and burial.

Landforms are traditionally taken as the result of geologically recent (late Cenozoic) activity, although landforms dating from the Mesozoic and even Paleozoic are increasingly well recognized especially from the Gondwanan continents. Early Permian landscapes from the upland Uncompahgre and Wichita systems were preserved as a result of active regional subsidence in Early Permian time that affected both the uplifts and surrounding regions. Exhumation of these landscapes occurred in the Cenozoic, associated with Laramide and more recent orogenesis and auxiliary drainage evolution in Colorado, and the distal effects of the Rio Grande Rift that extended to Oklahoma. The Lower Permian post-orogenic strata should thicken toward the core ARM highlands, along a trend perpendicular to that of the gravity anomaly and inferred mass load.

Conclusions

Geologic data from the core ARM system both long known and newly documented, indicate preservation of Early Permian landscapes that exhibit paleo-relief of as much as 1000 m, and record subsidence extending over a length scale of nearly 1500 km. In addition, geophysical data buttressed by geological data reveal a regional-scale mafic load underpinning these same regions. The correspondence of the gravity data with direct observation of high-density Cambrian mafic intrusives from the Wichita Mountains (Oklahoma) and Wet Mountains (Colorado) indicates that this signal relates to the formation of the early Paleozoic SOA. Geodynamic modeling of the effects of such a load in the presence of a horizontal stress field, such as that implicated in Pennsylvanian–Permian ARM orogenesis, indicates that the amplitude of flexurally supported features is modulated nonlinearly. This leads to buckling and thrust formation with the application of sufficient compressive stress, and subsidence of topography formed by buckling upon relaxation of the high compressional stresses. These results are used to suggest that the core highlands of the Ancestral Rocky Mountains, uplifted in Pennsylvanian time, ceased to rise and ultimately succumbed to load-induced subsidence in Early Permian time, spatially associated with high-density bodies in the upper crust. Unlike orogenic collapse, this phenomenon formed unassociated with any significant upper crustal structural or magmatic activity. Like orogenic collapse, however, this subsidence likely reflects readjustment of horizontal stresses. The termination of ARM deformation was related to stress release associated with closure of the Marathon segment of the Ouachita orogenic belt.

This shift in the regional stress field and associated cessation of northeast-oriented compressive stresses precipitated the geologically abrupt reversal from orogenic uplift to epeirogenic subsidence tied to the existence of an upper crustal mafic load, as documented in our data set and modeling. These results underscore the roles of inheritance and crustal inhomogeneity in erecting and ultimately eradicating a classically enigmatic intra-plate orogenic system.

Chapter 4. Paleogeography of the Unita-Piceance Basin Region of Colorado and northeastern Utah

Introduction

The rectangular Uinta-Piceance basin region of northwestern Colorado and northeastern Utah encompasses the Laramide (latest Cretaceous to Paleogene) Uinta and Piceance basins and surrounding uplifts. The area is mostly underlain by Phanerozoic sedimentary rocks that represent at least five different phases of basin evolution. Pennsylvanian and lower Wolfcampian rocks were deposited during one of these phases, generally referred to as the Ancestral Rocky Mountain orogeny. During this orogeny, the Uinta-Piceance region included parts of four major sedimentary provinces: the Eagle basin, the Paradox basin, the Wyoming shelf, and the Oquirrh basin. Episodic deposition also occurred on the Emery uplift and the Callville shelf. Pennsylvanian and lower Wolfcampian rocks are widespread in the subsurface of the Uinta-Piceance region, but outcrops are restricted to a few areas (**Figure 16**).

Recent studies demonstrate that Pennsylvanian and lower Wolfcampian strata of the Uinta-Piceance region are commonly cyclic, characterized by repetitive sequences of lithofacies that represent repeating successions of depositional environments. Comparable cyclic deposition in upper Paleozoic strata has been widely recognized globally and is generally attributed to eustatic fluctuations and associated climate changes forced by expansions and contractions of Gondwana continental ice sheets. Existing paleogeographic maps and reconstructions for the Uinta-Piceance region greatly homogenize geologic history by showing only the dominant lithology deposited during a particular cycle or sequence and not the variations in depositional environments and lithofacies represented in the sedimentary cycles. Inferred differences between eustatic transgressive and regressive deposition differ significantly from earlier paleogeographic reconstructions. Paired maps for four time intervals (Morrowan and early Atokan, late Atokan and Desmoinesian, Virgilian and Missourian, early Wolfcampian) are presented. These maps represent the duration of the Ancestral Rocky Mountain orogeny and also correspond to the Absaroka I sequence.

The Ancestral Front Range and Uncompahgre uplifts are the major late Paleozoic orogenic highlands in the Uinta-Piceance basin region. Subsiding areas include the Eagle, Paradox, and Oquirrh basins, the Wyoming shelf, the Emery uplift, and the Callville shelf. The Ancestral Front Range and Uncompahgre uplifts, cored by Precambrian crystalline rocks, form the eastern and southwestern boundaries of the Eagle basin, respectively. These uplifts had considerable structural and topographic relief and were bounded by narrow complex fault zones.

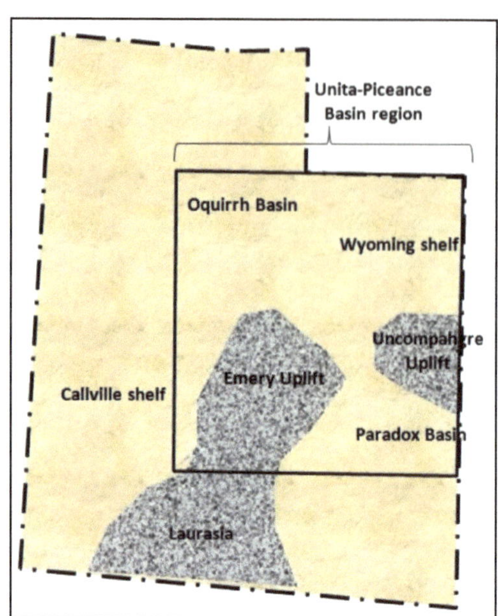

Figure 16. The Uinta-Piceance Basin region discussed in this section consists of the Wyoming shelf in the northeast corner, the Oquirrh Basin in the northwest corner, the Emery Uplift in the southwest corner, the Paradox Basin in the southeast corner, and Uncompahgre uplift in the east central part of Utah.

The Ancestral Sawatch uplift formed an intra-basinal block in the southern part of Eagle basin and was probably characterized by lesser positive relief during the late Paleozoic. The Emery uplift of central Utah was intermittently emergent during the Pennsylvanian to Wolfcampian. At those times it formed the northwestern boundary of the Paradox basin. The Emery uplift was never deeply eroded (Mississippian strata are preserved below an unconformity with Wolfcampian rocks), however, and thus probably had minimal topographic relief.

The Gore fault zone forms the boundary between the Ancestral Front Range uplift and the Eagle basin. Reactivation of this fault at least three times during the Mesozoic and Cenozoic has obscured its late Paleozoic history. The Gore fault is now locally a reverse fault. It was a normal fault during the late Paleozoic. The normal-fault hypothesis suggests a local extensional environment that is consistent with the presence of interbedded volcanic rocks in the Middle Pennsylvanian Minturn Formation within a kilometer of the Gore fault. Alternatively, the presence of a deep late Paleozoic trough adjacent to the Ancestral Front Range could reflect crustal loading from the east and reverse motion on the Gore fault.

The unexposed Garmesa fault zone forms the boundary between the Ancestral Uncompahgre uplift and the Eagle basin. The Garmesa fault was a late Paleozoic oblique-slip fault; but may have been a reverse fault in the late Paleozoic. Numerous intra-basinal faults (with inferred normal, reverse, and oblique-slip displacement) were also active in the Eagle basin during the late Paleozoic. These structures had significant influence on patterns of deposition and subsidence.

The northern part of the Paradox basin, which contains as much as about 2,400-2,700 m of Pennsylvanian and lower Wolfcampian strata also is in the Uinta-Piceance region. The northern Paradox basin is bounded on the northeast by the Uncompahgre fault and the Ancestral Uncompahgre uplift. Based on borehole, seismic, and structural data, the Uncompahgre fault was a reverse fault in the late Paleozoic. At least 9.6 km of horizontal displacement, and about 6 km of vertical displacement along the northern segment of the Uncompahgre fault occurred. The deepest part of the Paradox basin was adjacent to the Uncompahgre uplift, and the basin's marked asymmetry is consistent with thrust-induced flexural loading along its northeastern margin. The Uncompahgre uplift most likely plunges and dies out to the northwest in a complex structural zone characterized by thrust and reverse faults.

The Emery uplift or platform formed a low-relief shelf on the west flank of the Paradox basin and may (at least in part) represent a flexural bulge or "fore-bulge" coupled to the over-thrusted Uncompahgre uplift. The eastern margin of the Emery shelf is fault controlled (the Emery fault). Northwest-trending reverse faults on the northern flank of the shelf were described. The vertical displacement on these structures is probably not more than a few thousand feet. Permian strata of the northern Paradox basin postdate this postulated faulting and on lap the eastern flank of the Emery uplift. The northern Paradox basin is similar to the Eagle basin in that intra-basinal faults and lineaments had significant effects on depositional patterns and local rates of subsidence.

The Emery uplift is bounded on the west by a structurally complex zone characterized by Cretaceous thrust faults (Sevier orogeny) and Neogene extensional faults (Basin and Range). Pennsylvanian and Wolfcampian strata in this province are considered part of the Callville shelf depositional province. Callville shelf deposits probably on lapped the western flank of the low-relief Emery uplift, but this relationship is not clear because of the lack of outcrops and boreholes and because upper Paleozoic strata in this area apparently have been tectonically displaced 50-100 km eastward from their depositional site.

The southeastern part of the Oquirrh basin is in the northwestern part of the Uinta-Piceance basin region and contains as much as 7,100 m of Pennsylvanian and Wolfcampian strata. Basinal strata have been disrupted by Cretaceous thrust faults and Neogene extensional faults. Net eastward tectonic transport of 30-100 km for Oquirrh basin strata in the Uinta-Piceance region was projected.

Alternatively, in the area west of the Wasatch fault (the easternmost fault in the extensional Basin and Range province), the younger extensional deformation may have fortuitously cancelled out older Sevier shortening such that late Paleozoic spatial relationships remained relatively intact. The Oquirrh basin was a northwest-trending trough that had its maximum topographic relief in the Late Pennsylvanian and Early Permian. The bounding faults of this trough are apparently masked by younger sedimentary rocks and structural features and have not been positively identified. The Leamington lineament represents the southern boundary of the Oquirrh basin, and the "Oquirrh-Uinta arch" partly defined the northern margin of the deepest part of the basin. The nature of the boundaries between the Oquirrh basin and the Paradox basin and the Wyoming shelf to the southeast and east are obscured by structural complexities and (or) burial under thick Mesozoic and Cenozoic strata.

In summary, the Uinta-Piceance basin region is characterized by a complex mix of compressional and extensional tectonic styles from the Early Pennsylvanian through the Early Permian. This mix of structural styles is most consistent with regional strike-slip deformation probably driven by interactions along distant continental margins. The relative importance of each continental margin as a driving force for intra-plate deformation is explored from the perspective of subsidence analysis later.

Late Paleozoic Climate, Eustacy, and Cyclicity

The late Paleozoic was the time of the last major pre-Quaternary glaciation. Initiation of this glacial age is generally correlated with the closure of several ocean basins and the resultant amalgamation of continental landmasses in the Southern Hemisphere to form the western half of Pangea. During this period, North America was rotated clockwise about 30°-40° from its present orientation, and the Uinta-Piceance basin region was at a latitude ranging from about 0°-5° N in the Early Pennsylvanian to about 15° N in the Early Permian. Extensive eolianites and evaporites in upper Paleozoic deposits of the Uinta-Piceance region indicate an arid climate. Paleo-wind directions were primarily to the west (to the south in present-day coordinates). Using the Quaternary as an analogue, numerous authors have suggested that the late Paleozoic ice sheets expanded and contracted at fairly regular time intervals, probably forced by Milankovitch orbital parameters.

As in the Quaternary, cyclic eustatic and climatic changes are the inferred effects of these ice-sheet fluctuations. Late Paleozoic sea-level curves are currently being refined. Sea level changes are generally interpreted as asymmetric (similar to Quaternary analogues), characterized by rapid transgressions (rapid melting of ice sheets) and slow regressions (slow buildup of ice sheets). The magnitude of Pennsylvanian cyclic eustatic changes ranged from tens of meters to almost 200 m suggesting an average sea-level fluctuation of 90 m. Coastal on lap curves indicate that eustatic fluctuations were superimposed on a significant global sea-level rise from the Morrowan to the end of the Desmoinesian. The magnitude of coastal on lap remained approximately the same in the Missourian, Virgilian, and Wolfcampian.

Deposition during late Paleozoic transgressions and regressions produced sedimentary "cycles" or depositional sequences that have been recognized and correlated across basins and globally. Many of these depositional sequences are not "cyclic" in the strict sense in that preferred ordering of lithofacies may not be present. This lack of regular ordering within a transgressive- regressive sequence can reflect the variable effects of tectonic activity, sediment supply, basin geometry, and other factors. The term "cycle" is used in the more general way to describe a transgressive-regressive depositional sequence. The periodicity of late Paleozoic sedimentary cycles or depositional sequences in the North American midcontinent outcrop belt was approximately 235,000-400,000 years, matching the Milankovitch eccentricity parameter. Based on a synthesis of 19 studies of late Paleozoic cycles, a typical periodicity of 1.3 million years, significantly longer than inferred late Cenozoic eccentricity was estimated, suggesting that a longer Milankovitch parameter may have been active in the late Paleozoic. Late Paleozoic transgressive-regressive sequences averaged about 2 million years (range of 1.2-4 m.y.) but also implied that these sequences could include four or five cyclothems.

The Milankovitch hypothesis was challenged, suggesting that eustatic changes may have been controlled by other processes including tectonism and large-scale melt-water discharge. A mechanism for repetition of these inferred alternative cycle-controlling mechanisms was considered. Until such a mechanism has been described and justified, Milankovitch orbital parameters of unknown periodicity provide the most compelling rationale for the repetitive late Paleozoic eustatic and climatic changes.

Late Paleozoic transgressive-regressive cycles or sequences, primarily reflecting the eustatic controls discussed above, have been reported from most of the well described stratigraphic units in the Uinta-Piceance basin region. Because much of the basinal topography in the Uinta-Piceance region was at or near sea level in the late Paleozoic, inferred eustatic fluctuations significantly affected the location of the marine-nonmarine interface and other depositional patterns. Changes in these depositional patterns representing deposition during maximum transgression and regression are presented in the following figures. In one case (late Atokan-Desmoinesian), regression limits of an evaporite facies in the Eagle basin was deposited during but not at the peak of regressions because of the facies importance in understanding regional depositional patterns. By presenting paired maps this dominant lithofacies or environment may represent deposition during a transgression, a regression, or an intermediate stage.

The time intervals selected for the maps (Morrowan and early Atokan, late Atokan and Desmoinesian, Missourian and Virgilian, early Wolfcampian) coincide with inferred major changes in regional tectonic and (or) depositional patterns. The maps are preliminary models, inhibited by the general lack of outcrop of late Paleozoic rocks and (or) the lack of detailed sedimentologic studies. Reliance on geophysical or lithologic (AmStrat) borehole logs was attempted for reconstruction of paleogeography in northwestern most Colorado and the Paradox basin-Emery uplift area. These borehole logs generally lack well-constrained age data and systematic lithologic descriptions.

Deposition in areas where Pennsylvanian rocks are no longer present is represented if there is compelling evidence that the rocks were originally deposited and later eroded.

Essentially no data are available from the subsurface of the Paleogene Uinta basin, and the inferred facies patterns in this important area where the Paradox basin merges with the Wyoming shelf and the Oquirrh basin are particularly speculative. Finally, due to gross uncertainties an attempt to palinspastically restore strata of the Oquirrh basin and Callville shelf to their depositional sites (pre-Mesozoic shortening and Cenozoic extension) was not conducted. Future studies will undoubtedly refine estimates of the amounts of tectonic transport and make accurate palinspastic restorations possible. A discussion of the Uinta Basin subsurface geophysical investigations is presented later in this chapter.

Sub-Pennsylvanian Geology

Morrowan strata unconformably overlie Mississippian rocks throughout most of the Uinta-Piceance region. In the Eagle basin and most of northwestern Colorado, Morrowan strata generally overlie the Lower Mississippian Leadville or Madison Limestone. Pennsylvanian strata (including the latest Mississippian (?) to early Middle Pennsylvanian Molas Formation) overlie the Leadville Limestone, the Doughnut Formation, or the Meramecian Redwall Limestone in the northern Paradox basin area, and the Redwall Limestone underlies Wolfcampian strata on the Emery uplift.

The contact between the Upper Mississippian Doughnut Formation and the Lower Pennsylvanian Round Valley Limestone in the Wyoming shelf area is problematic. Sedimentation was continuous or only slightly interrupted from Chesterian (latest Mississippian) to Morrowan time. The Round Valley Limestone represents a marine invasion over the deltaic and estuarine environments of the Chesterian Doughnut Formation. Near the Utah-Colorado border, however, some post-Doughnut, pre-Morrowan erosion clearly occurred because the Morgan Formation (in part laterally equivalent to the Round Valley) rests unconformably on the Madison Limestone and the Doughnut Formation is missing. This unconformity may pass westward into an unconformity along the Wyoming shelf. To the west, in the Oquirrh basin and on the Callville shelf, Morrowan strata are inferred to conformably overlie Chesterian rocks. Isopach maps indicate that the Oquirrh basin and Wyoming shelf began to subside as discrete tectonic elements in the Late Mississippian, significantly before subsidence began in the Eagle and Paradox basins.

Early Middle Pennsylvanian
Morrowan and Early Atokan

Within the Pennsylvanian and Wolfcampian strata of the Uinta-Piceance region, depositional cycles or sequences are least recognized in the generally thin sections of Morrowan and lower Atokan strata. This relative lack of recognition probably reflects several factors, including poor outcrops, little recent detailed sedimentologic analysis, slow subsidence rates, and eustatic fluctuations of comparatively lesser magnitude. Ten (10) significant eustatic fluctuations in the Morrowan and lower half of the Atokan that had inferred magnitudes of from 20 to about 80 m, about half the size of the magnitude inferred for late Atokan through Early Permian eustatic fluctuations. Cycles from a 265-m-thick section of the Callville Limestone on the Callville shelf (Route 148 southwest section) is an example (**Figures 17 & 18**).

These cycles consist of a lower fine-grained sandstone overlain by an upper thin-bedded limestone. Farther south in the Cove Fort area, Morrowan strata (255 m thick) are mainly dolomitic limestone but include some sandstone and oolitic limestone. This suggests that cyclic sedimentation in this area reflects repeated transgression (carbonate platform deposition) and regression (shelf sand deposition) over an unstable border between a shelf to the east and a basin to the west.

Figure 17. The Cedar Breaks National Monument exposes the Callville Limestone belonging to the Calville shelf . UT 148 follows the rim of the shelf with many turn off overlooks into the eroded canyons produced by weathering carbonates.

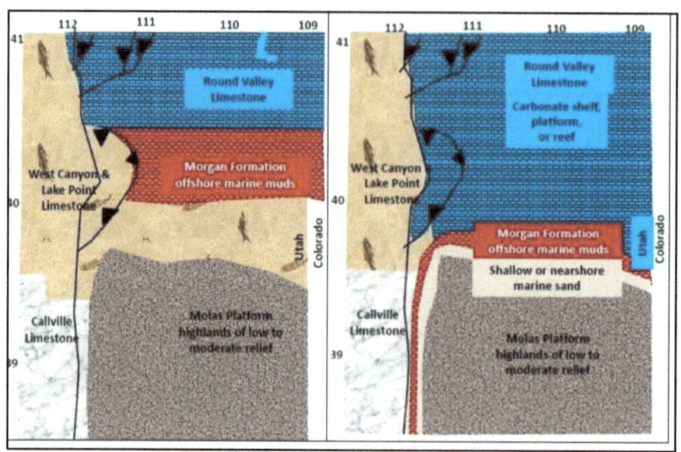

Figure 18. Maps showing the Morrowan to early Atokan paleogeography of the Uinta-Piceance basin region during maximum regression (left) and transgression (right). The vertical line on the left side of the maps represents the Wasatch Fault.

A thin Atokan sequence of the Callville Limestone in the Callville shelf area overlain by an unconformity was noted. Desmoinesian strata were deposited across the Callville shelf but subsequently eroded. Cycles from the Atokan part of the Callville from the Route 148 southwest section consists of a basal gray to brown shale overlain by a middle unit of gray to yellowish-gray sandstone and an upper unit of thin-bedded limestone or dolomite. Similar sandstone is cross bedded in the Cove Fort section. The limestone was not described, but it probably represents deeper facies (transgressive deposits) deposited at the beginning of cycles whereas the sandstone forms the upper parts of cycles and was deposited during regressions.

In the Oquirrh Mountains of the southeastern Oquirrh basin, the Lake Point Limestone and the West Canyon Limestone include about 365-440 m of mainly Morrowan strata. Cycles in the Lake Point Limestone that comprise were described in ascending order as shaly limestone, clastic calcareous sandstone, calcitic arenaceous limestone, cross bedded arenaceous limestone, cherty limestone, and bedded chert. The West Canyon Limestone, comprising arenaceous limestone, cherty argillaceous limestone, and cross bedded sandstone is also characterized by cyclic deposition but no description of a typical cycle were completed. These cycles resemble those described for the Callville shelf and have an inferred similar origin (**Figure 19**).

Detailed petrographic descriptions of a section 321 m thick of similar age and lithology at Cascade Mountain in the central Wasatch Range were not identified cycles but interpreted as a variety of shallow-marine depositional environments that changed in response to variations in sea level. A significant change in lithology and depositional style was described in the southeastern Oquirrh basin between Morrowan strata and the more sandstone-rich Atokan to Desmoinesian Butterfield Peaks and Erda Formations.

The timing of this change is similar to that noted for the Callville shelf. These changes from late Atokan to early Desmoinesian were probably initiated in the early to middle Atokan (**Figure 20**).

Figure 19. The Bingham Canyon Mine exposes the strata of the Lake Canyon and West Canyon Limestones of Morrowan age in the Oquirrh Mountains on the southeastern side of the Oquirrh Basin. The mine exposes the Bingham Canyon Formation.

The Emery uplift, on which Permian rocks unconformably overlie Mississippian strata, bounds the Callville shelf to the east. Strata on the Callville shelf are relatively fine grained, and it is possible that Morrowan and lower Atokan strata were deposited over at least the westernmost part of the Emery uplift and subsequently eroded. The area that now includes the eastern part of the Emery uplift, the Paradox basin, and at least part of the Ancestral Uncompahgre uplift was probably emergent or partly emergent. Latest Mississippian (?) through early Atokan deposition in the Paradox basin is recorded by the Molas Formation, mainly a red regolith that formed by weathering of Mississippian limestone. The emergent area is here informally named the "Molas platform".

.

Figure 20. Cascade Mountain in the Wasatch Range represent shallow-marine depositional environments that changed in response to variations in sea level were not cyclical environments as in the Oquirrh Mountains.

Morrowan and lower Atokan strata of the Wyoming shelf depositional province are generally less than about 150 m thick and include the Round Valley Limestone and the lowest part of the Morgan Formation. The Round Valley Limestone crops out along the south and north flanks of the Uinta Mountains and extends east as far as Juniper Mountain. It consists mainly of thin-bedded to massive limestone, commonly dolomitic or cherty. Siliciclastic deposits, generally gray or purplish-red mudstone are typically a minor component but locally (for example, Sols Canyon) comprise as much as 35 percent of Round Valley sections. A significant clastic component to the Round Valley Limestone (mapped as the lower member of the Morgan Formation) in the Duchesne River area is present (**Figure 21**).

The proportion of clastic material in Morrowan and lower Atokan strata (or strata that overlie Mississippian rocks and are of inferred Morrowan and early Atokan age) increases in the subsurface south and southeast of the Uinta Mountains outcrop belt in strata assigned to the Morgan Formation. This increase suggests that the "Molas platform" contributed clastic detritus to the Wyoming shelf. The more abundant limestone in this province was probably deposited during transgressions, when clastic detritus was trapped closer to its source. Regressions were probably characterized by lowering of base level and more widespread clastic deposition.

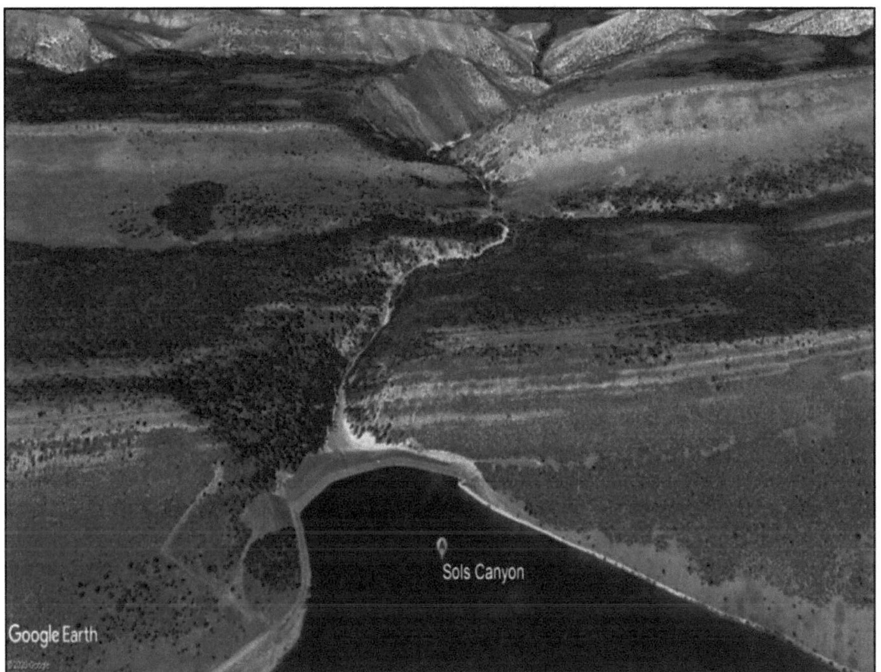

Figure 21. Sols Canyon represents part of the Wyoming shelf Round Valley Limestone with minor clastics present.

The Morrowan and lower Atokan Belden Formation (as thick as 200 m or more) of the Eagle basin represents deltaic and prodeltaic deposition off the Ancestral Front Range. The best exposures of cyclic deltaic deposits are in the Sweetwater Creek area about 30 km northeast of Glenwood Springs.

Typical cycles coarsen upward from (1) prodeltaic deposits of gray to black shale, dolomite, and limestone to (2) delta-front deposits of wave-rippled, fine- to coarse grained sandstone and interbedded dark mudstone. Some cycles are capped by fluvial channel deposits of trough cross bedded, coarse-grained to conglomeratic sandstone. Paleo-current data indicate that sediment transport was to the southwest. The texture and sedimentology of these deltaic rocks indicate contemporaneous uplift in the Ancestral Front Range. A large delta system in the overlying Minturn Formation in the McCoy area was present in the same location as the large Belden delta. The large Minturn delta was fed by a major drainage system channeled through a west-trending transfer zone in the Gore fault.

Throughout most of Eagle basin, the Belden Formation consists of interbedded gray limestone and dark-gray, commonly calcareous, mudstone. These mudstones are the inferred regressive deposits and represent maximum progradation of deltas and associated prodelta clastic sediments. During transgressions, clastic sediments were trapped closer to uplifted sources and limestones were deposited over broader areas.

Beds of anhydrite and (or) gypsum in the Belden Formation have been found in the subsurface on the north flank of the White River uplift, and in outcrops on the west side of the Sawatch uplift. Interbedded facies (shale and limestone) at each location suggest a basinal setting. These evaporites are inferred as precipitated from basinal brines in fault-bounded sub-basins during eustatic low stands.

Evidence from an area southeast of the Uinta-Piceance region was presented that major coarse-grained delta systems in the Belden Formation had a source in the Sawatch uplift. The lack of comparable deltaic deposits on the west side of the Sawatch uplift in the Uinta-Piceance region suggests either that the uplift had markedly asymmetric relief or that it was significantly narrower than the modern Sawatch Range. Based on the dominance of limestone and dark mudstone in the Belden Formation in the Crested Butte area, it was concluded that the Ancestral Uncompahgre uplift had not developed in the Morrowan and early Atokan. The "Molas platform" apparently did not shed significant sediment eastward into the Eagle basin. The eastern boundary of this low-relief source terrane was probably west of the Middle and Late Pennsylvanian margin of the Uncompahgre uplift.

Due to a lack of subsurface data, the location of the Morrowan-early Atokan structural boundary between the Front Range and the Eagle basin is not well constrained. North and northwest of the McCoy area, coarse grained deltaic deposits have not been recognized adjacent to the basin margin either in outcrops or in the subsurface. Morrowan and lower Atokan strata are probably not present where a thin (120 m) Pennsylvanian section rests on Mississippian rocks. It may be that Morrowan and lower Atokan rocks were never deposited eastward. Alternatively, Morrowan and lower Atokan rocks could have been deposited easterly and then eroded later in the Pennsylvanian.

Summarizing, Morrowan and lower Atokan strata across most of the Uinta-Piceance basin area consist of alternating fine-grained clastic rocks (inferred regressive deposits) and more abundant limestone (transgressive deposits). Significant Morrowan and early Atokan tectonic uplifts include the Ancestral Front Range and Sawatch uplifts.

Middle and Late Middle Pennsylvanian
Late Atokan and Desmoinesian

The late Atokan and Desmoinesian paleogeographic maps highlight three major changes that occurred during the early and (or) middle part of the Atokan in the Uinta-Piceance region: (1) uplift of the Ancestral Uncompahgre uplift and the Emery uplift; *(2)* subsidence of the Paradox basin; and (3) significant progradation of clastic sediments across the Wyoming shelf. At least 24 major eustatic fluctuations during the last half of the Atokan and Desmoinesian had inferred magnitudes of as much as 200 m. As described earlier, upper Atokan and Desmoinesian strata probably have been eroded from the Callville shelf. Accordingly, the early Atokan depositional pattern (described above) continued into the late Atokan and Desmoinesian.

The increased clastic detritus on the Callville shelf was derived from both the Uncompahgre and Emery uplifts.

The Butterfield Peaks and Erda Formations represent Atokan and Desmoinesian deposition in the Oquirrh basin. From detailed sedimentologic studies in the Mount Timpanagos area on the upper plate of the Charleston-Nebo thrust and in the southeastern Oquirrh basin, lithofacies were deposited in environments ranging from normal-marine carbonate shelf to eolian dunes, inferred to represent deposition during maximum transgressions and regressions, respectively (**Fig**).

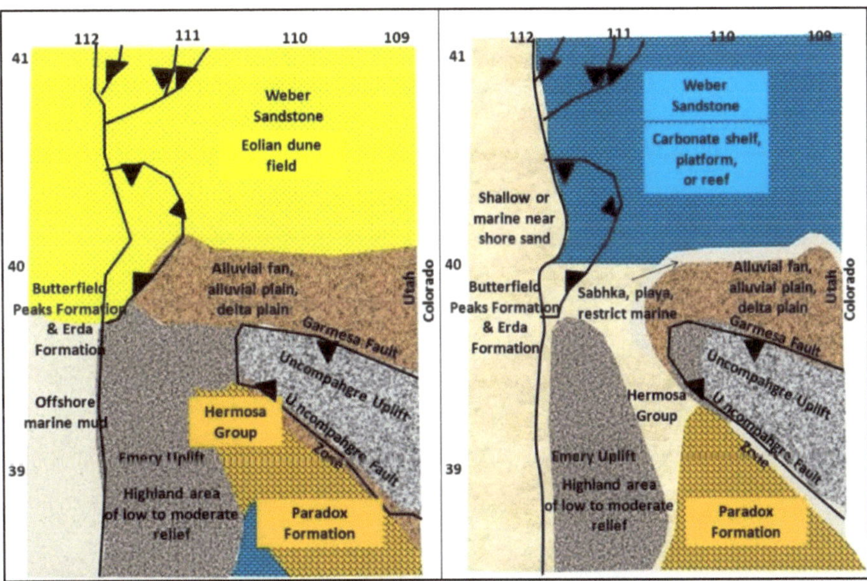

Figure 22. Maps showing the late Atokan to Desmoinesian paleogeography of the Uinta-Piceance basin region during maximum regression (left) and transgression (right).

Statistical analysis suggests that these rocks are characterized by several preferred vertical facies transitions but were not cyclic in the strict sense. Suggested factors that might have obscured better development of cycles include (1) rapid, possibly episodic subsidence, (2) rapid deposition of windblown sand and silt, and (3) rapid deposition of carbonate rocks in environments not overwhelmed by sand. Eolian deposits are much less common in the Oquirrh Group west of the Wasatch fault than they are at Mount Timpanagos. Shoal and emergent carbonate deposits are present at several locations west of the Wasatch fault.

These deposits, along with too numerous alternations of sandy and carbonate facies, show that there is not a one-to-one correspondence between lithology and transgressive-regressive depositional patterns. Similar lithofacies were described in the Oquirrh Group in the Oquirrh Mountains and at several other locations west of the Wasatch fault. A number of authors suggested that these strata are cyclic, mainly on the basis of the obvious alternation of siliciclastic and carbonate sediments.

Similar Middle Pennsylvanian rocks were described immediately west of the Uinta-Piceance basin region. They described an idealized transgressive-regressive facies sequence within the carbonate strata, bounded by siliciclastic units that represent the most regressive facies. Based on extensive paleo-current measurements, sand transport in the Oquirrh basin during the Atokan and Desmoinesian was to the south. In addition, sandstone deposition was slightly more dominant in the northern and eastern parts of the basin than in the southern and western parts.

The sands probably were derived from the craton to the north and northeast and transported to the Oquirrh basin primarily via the Wyoming shelf, and significant sand input from either the Emery or Uncompahgre uplift seems unlikely. Mean composition of siliciclastic grains (2-3 phi) is within the quartz-rich end of the subarkose field, a mature composition for this grain size. It was pointed out that because the Oquirrh basin subarkosic sands (Late Pennsylvanian and Early Permian) and Pennsylvanian subarkosic sands derived from crystalline rocks in the Front Range uplift are compositionally similar, the source of sands in the Oquirrh basin may have been the crystalline rocks of the nearby Uncompahgre uplift.

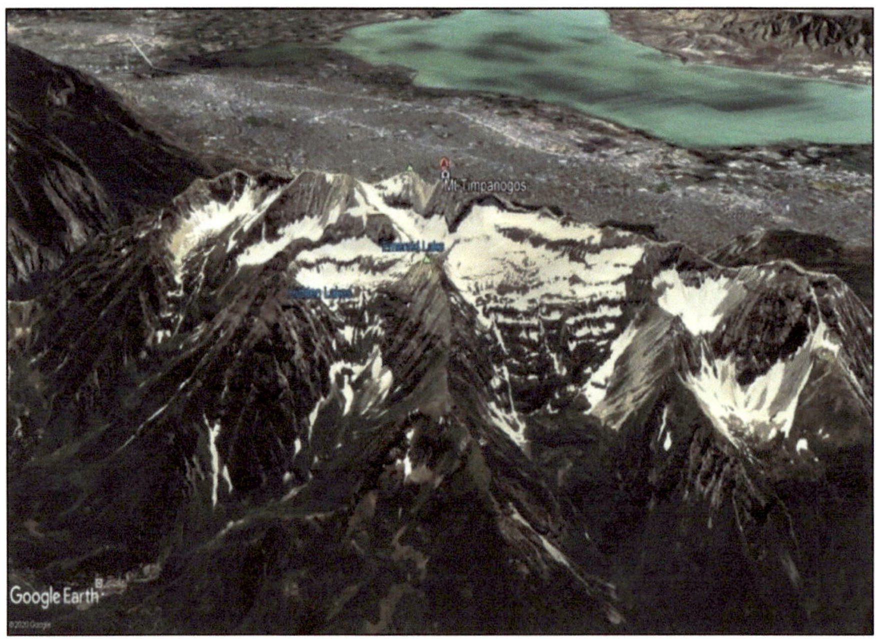

Figure 23. Mount Timpanogos rests on the upper plate of the Charleston-Nebo thrust, lithofacies were deposited in environments ranging from normal-marine carbonate shelf to eolian dunes, inferred to represent deposition during maximum transgressions and regressions, respectively.

The Emery uplift was emergent, but preservation of Mississippian rocks below a Permian unconformity indicates that this uplift had minimal relief, probably was not a major sediment source, and may have been on lapped significantly during maximum transgressions (**Figure 223**)

A delta system derived from the Uncompahgre uplift fed the Oquirrh basin during the Middle and Late Pennsylvanian, but this hypothesis is difficult to assess because neither outcrops nor subsurface data are available in the critical area between the Uncompahgre uplift and the Oquirrh basin and fluvial-deltaic strata have not been recognized in the Oquirrh basin. The presence of a complex structural zone on the northwest flank of the Uncompahgre uplift that might have served to channel Uncompahgre detritus to the northwest was documented. If Uncompahgre-derived sand did enter the southern part of the Oquirrh basin then it probably was reworked and transported south onto the Callville shelf. Sedimentary cycles in the upper Atokan and Desmoinesian upper member of the Morgan Formation of the Wyoming shelf depositional province were described. These cycles typically pass upward from (1) carbonate rocks deposited in normal-marine shelf environments (maximum transgression deposits), to (2) carbonate rocks deposited in restricted-shelf and near shore environments, to (3) sandstones deposited in eolian dunes (maximum regression deposits).

Sands were derived from the north across the Wyoming shelf. Transgressive-regressive depositional patterns on the Wyoming shelf and in the Oquirrh basin (and by inference, on the Callville shelf) were thus fairly similar. Atokan-Desmoinesian strata on the Wyoming shelf (Morgan Formation and lower part of Weber Sandstone) have a maximum thickness of about 460 m whereas correlative strata in the Oquirrh basin (the Butterfield Peaks Formation) are as thick as 2,765 m. Because sediment supply and subsidence were in balance in the Oquirrh basin, the basin was not characterized by significant negative topographic relief at this time. If these two areas experienced the same number of transgressive-regressive events during this period, then the Oquirrh basin depositional cycles should be much thicker than those that formed on the Wyoming shelf.

The Paradox basin formed as a discrete tectonic element in the Atokan by fragmentation of the "Molas platform." The low-relief Emery uplift (discussed above) and the Uncompahgre uplift (of inferred high relief) formed the western and east-northeastern margins of the Paradox basin, respectively. Erosion of Paleozoic cover rocks from the Uncompahgre uplift exposed Precambrian crystalline rocks considered to be the main source of clastic detritus in the Paradox basin.

As described earlier, seismic data show that the late Paleozoic boundary between the northern Paradox basin and the Uncompahgre uplift was a reverse fault, and basin asymmetry is inferred to reflect flexural loading along the northeastern margin of the basin.

A maximum thickness of about 1,700 m for the Atokan and Desmoinesian in the northern Paradox basin including strata of the Pinkerton Trail and Paradox Formations of the Hermosa Group and undivided rocks of the Hermosa Formation. The Atokan to lower Desmoinesian Pinkerton Trail Formation is the lowest unit and consists of mixed carbonate and fine-grained clastic rocks.

The Pinkerton Trail is thickest on the western margin of the basin and thins toward the center of the basin where it inter-tongues with and is overlain by the Paradox Formation which consists of evaporites, black shale, and limestone. The undivided rocks of the Hermosa Formation crop out on the northeastern margin of the basin and are coarse-grained clastic rocks derived from the Uncompahgre uplift.

Atokan and Desmoinesian sedimentary cycles in the Paradox basin are well described. At least 29 evaporite cycles were identified on a Paradox basin cross section from borehole data. These basinal evaporite cycles ideally consist of highly calcareous, organic-rich black shale (inferred maximum transgression deposits), anhydrite, dolomite and silty dolomite (intermediate sea-level stand), and halite and potash salt deposits (maximum regression). Some halite beds exhibit evidence of subaerial exposure and erosion. On the western margin of the basin cycles were described that ideally consist of calcareous black shale (inferred maximum transgression deposits), dolomite and siltstone (intermediate sea level), and algal and fossiliferous limestone (maximum regression).

Strata north of the evaporite limit in the northern most Paradox basin consist mainly of interbedded fossiliferous limestone (inferred transgressive deposits), and variegated fine grained clastic deposits of inferred sabkha or playa origin (inferred regression deposits) (**Figure 24**).

It should be noted that the basinal and western-basin margin cycles of the Paradox described above are anomalous in comparison to other regionally correlative cycles in that the most clastic rich intervals have been assigned to transgressive phases (that is, raised base level) whereas clastic sediments elsewhere are inferred to have been trapped in basin-margin settings. The clastic component in the Paradox basin reflects reworking of basin-margin clastic sediments by transgressing seas. Perhaps a more suitable explanation involves sediment accumulation rates. The thick halite intervals in the cycles may have deposited at rates of 5 cm per year or more whereas the highly calcareous, organic-rich, black shales probably accumulated at rates of a few millimeters per year or much less. Given these rates, it may be that input of clastic sediment to the basin actually increased during regressions, but that this increase is masked by the great volume of the rapidly accumulating evaporites. The evaporites were deposited subaqueously in a deep basin through precipitation from a concentrated brine, an interpretation that requires a basin with restricted circulation.

Figure 24. Strata north of the evaporite limit in the northern most Paradox basin consist mainly of interbedded fossiliferous limestone (inferred transgressive deposits), and variegated fine grained clastic deposits of inferred sabkha or playa origin (inferred regression deposits). Limestones are the grayish green to tan colored rocks. Clastics are the reds and browns. View is from Onion Valley Drive.

The Emery uplift was emergent and provided a partial barrier to circulation. Several possible inlets to the Paradox basin may have been present including (1) the "Oquirrh sag," a presumed connection with the Oquirrh basin in the structurally complex zone between the northwestern end of the Uncompahgre uplift and the northern Emery uplift, and (2) the "Fremont sag," a possible east-west connection between the Paradox basin and the Callville shelf to the south of the Uinta-Piceance region. Given the non-marine to marginal-marine regressive facies described above for the Oquirrh basin and Callville shelf, it is unlikely that either inlet was active during maximum regressions. The presence of fossiliferous limestone in the northernmost Paradox basin, however, requires normal or near-normal marine waters in the basin during transgressions, and this water must have entered the basin without mixing with the evaporitic brines. This non-mixing requirement is more easily met for the Oquirrh sag, and therefore it was an inlet into the Paradox basin during maximum transgressions. In the area of the Oquirrh sag anomalously thin sections (less than 220 m) of inferred but undated Pennsylvanian strata and do not provide sufficient data to test the hypothesis.

Tectonic activity was a major control on evaporite deposition in the Paradox basin. Based on apparent rates of subsidence in the Paradox basin, the Uncompahgre uplift probably experienced its highest rates of over thrusting, and extended farthest to the northwest, during the Desmoinesian. Decreased activity on the Uncompahgre would have resulted in decreased sedimentation rates (and accommodation space for evaporites), greater progradation of clastic sediments from basin margins, and widening of the Oquirrh sag, all effects detrimental to evaporite deposition and preservation.

The boundary between the Eagle basin and Wyoming shelf depositional provinces is gradational north of the Uncompahgre uplift near the Colorado-Utah State line. In general, varicolored clastic rocks become finer grained and are interbedded with greater proportions of limestone with increasing distance from the Uncompahgre uplift.

Anomalous limestone-poor strata in the subsurface indicate lateral lithologic variability and the possible presence of a north-flowing fluvial-deltaic system. During regressions, alluvial clastic sediments derived from the Uncompahgre probably merged northward with south prograding eolian deposits. The greater width of the belt of Uncompahgre-derived clastic sediments north of the uplift in Utah than in the Paradox and Eagle basins probably reflects the lower subsidence rates characteristic of the Wyoming shelf depositional province. In more rapidly subsiding areas, Uncompahgre-derived clastic sediments are inferred to have been trapped closer to basin margins.

During regressions, alluvial clastic sediments derived from the northern Front Range uplift probably inter-fingered with south-prograding dunes in the northeastern Eagle basin, but there is little subsurface data to document this relationship. The southern limit of the Wyoming shelf dune field probably was in the Juniper Mountain area. Thick beds of white and red sandstone, presumably eolian are present in the upper part of the Morgan Formation in and at Juniper Mountain. Thick beds of well-sorted sandstone are almost absent farther southeast in the subsurface.

Upper Atokan and Desmoinesian strata of the Eagle basin include the Minturn and Gothic Formations on the eastern and western margins of the basin, respectively, and the Eagle Valley Evaporite in the central part of the basin. Highly variable shoaling upward sequences in the Minturn Formation in the Minturn area were described. Maximum regression deposits are coarse alluvial and fan-delta clastic rocks; maximum transgression deposits are offshore marine mudstone and biostromal limestone. During transgressions, limestone was deposited within a few kilometers of the Gore fault. Relief was significant on the Front Range uplift, the source for the Minturn Formation, and minor on the Sawatch uplift, which he considered an influence on depositional patterns but not a significant sediment source. A large deltaic system in the McCoy area of Colorado was channeled through a west-trending transfer zone in the Gore fault (**Figure 25**).

Figure 25. The Minturn Formation is exposed in road cuts along Colorado Highway 131 at the town of Bond and northward to McCoy. The formation is also exposed west of McCoy.

A shoaling upward sequence similar to those in the Minturn Formation from the Gothic Formation in the Crested Butte and Crystal River areas, respectively was interpreted. It was concluded that both the Uncompahgre and Sawatch uplifts were significant sediment source terranes and that relief was greater on the Sawatch.

The main input of marine waters was input into the "Gothic trough" from the south. In the central part of Eagle basin (Eagle Valley location), shoaling-upward cycles in the Eagle Valley Evaporite were present. Micritic carbonates are inferred maximum transgression deposits. Because Desmoinesian subsidence rates were greater in the Eagle basin than on the Wyoming shelf, sea water was apparently trapped in the Eagle basin during regressions and gypsum was precipitated from an evaporative brine for much of the duration of the regressive phase. Although gypsum is widespread over the central part of Eagle basin, halite is present in only a few discrete areas. These areas of thick halite deposition have been interpreted as topographically low areas produced by syn-depositional faulting (**Figure 26**).

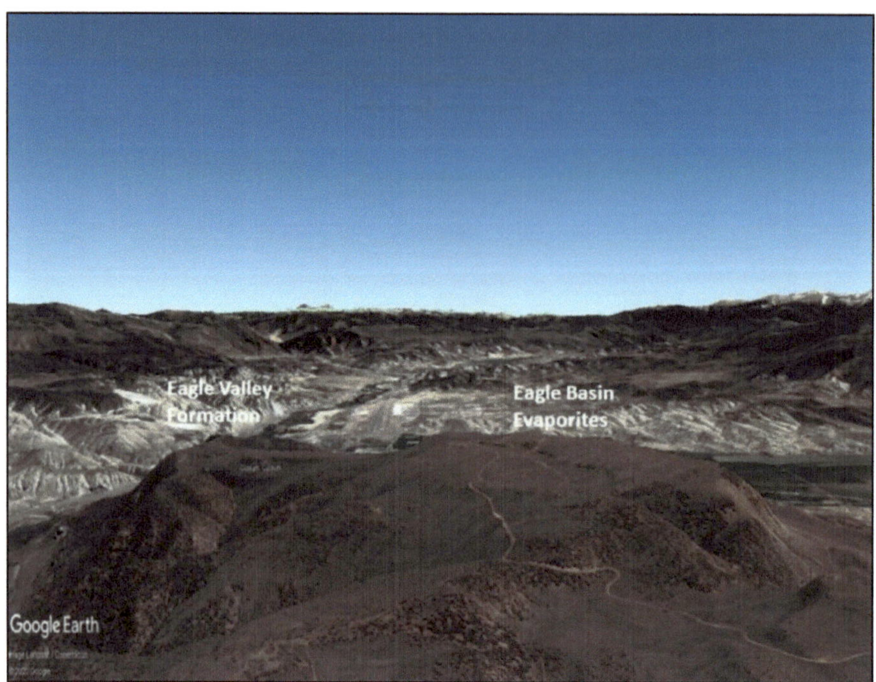

Figure 26. The Eagle Basin in Colorado exposes the Eagle Valley Formation and Eagle Basin Evaporites representing shoaling upward cycles. Seawater was trapped in the basin during regressive cycles allowing gypsum to evaporate from concentrated brine.

During maximum regressions, shallow marine to non-marine clastic sediments prograded over evaporitic deposits. Access to normal marine water was completely cut off to the north by Morgan eolianites. Consistent with the inferred paleogeography, the lack of a marine fauna in clastic rocks of the Eagle Valley Evaporite suggests that low stands (regressive phases) were characterized by lacustrine like circulation. The geometry and facies of clastic wedges at the tops of cycles in the Eagle Valley Evaporite indicate progradation from both the Uncompahgre and Front Range uplifts in early Atokan position. The very thin Pennsylvanian section in the subsurface closest to the Front Range uplift is consistent with a complex history of basin-margin faulting.

Summarizing, the late Atokan to Desmoinesian history of the Uinta-Piceance region is characterized by increased tectonic activity reflected in higher rates of basin subsidence and in continued or initial uplift and unroofing of the Front Range, Sawatch, and Uncompahgre uplifts. The combined effects of tectonism and eustasy led to restricted circulation in the Eagle and Paradox basins in which evaporite deposition was significant. Deposition during regressions was also characterized by significant progradation of eolian sands across the Wyoming shelf. Deposition of limestone dominated during transgressions.

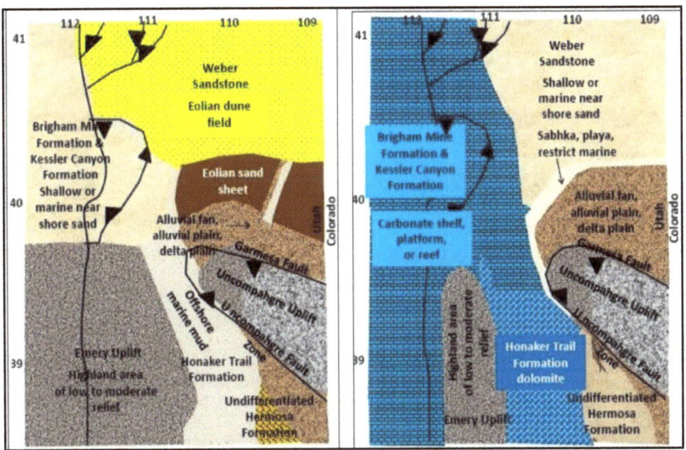

Figure 27. Paleogeography of the Missourian to Virgilian during maximum regression (left) and transgression (right).

Late Pennsylvanian Missourian and Virgilian

The major uplifts and basins of the Uinta-Piceance region continued to be active during the Missourian and Virgilian. Evaporite sedimentation essentially ended in the Paradox and Eagle basins, probably in response (1) a general rise in sea level leading to less restricted circulation, (2) decreasing rates of subsidence in evaporite basins, leading to increased progradation of clastic sediments derived from basin-margin uplifts, and (3) changes in tectonic geomorphology, leading to less topographic restriction of basins. It was suggested there were 17 major eustatic fluctuations during the Missourian and Virgilian, with magnitudes ranging from about 60 to 180 m (**Figure 27**). Alternatively, it was recognized that evidence for 36 major eustatic fluctuations in the North American midcontinent during the Missourian and the lower part of the Virgilian occurred.

Missourian-Virgilian rocks are not preserved on the Emery uplift. No Missourian and only a thin (<100 m) section of Virgilian limestone are present on the northern Callville shelf. These areas were characterized by minor uplift during most of this interval, but it also was suggested that Missourian and Virgilian strata may have been deposited and then subsequently removed during a latest Pennsylvanian erosional event. Neither hypothesis suggests more than minor subaerial or subaqueous relief in these areas, perhaps close to the inferred magnitude of eustatic fluctuations. Accordingly, emergence during maximum regressions and partial submergence and carbonate deposition during maximum transgressions may have occurred. Periodic emergence and erosion of the Callville shelf is consistent with the presence of clasts in correlative rocks in the adjacent Oquirrh basin that include fossils slightly older than their host strata.

Relatively rapid subsidence continued in the southeastern Oquirrh basin where Missourian and Virgilian strata as thick as 2,200 m include the Bingham Mine Formation and the lower part of the Kessler Canyon Formation. In contrast, strata of inferred Missourian and Virgilian age on the east flank of the Oquirrh basin on the Wyoming shelf (the middle part of the Weber Sandstone) are at most a few hundred meters thick. Both the Bingham Mine and Kessler Canyon Formations consist of sandstone and lesser amounts of interbedded limestone. Sandstone is typically massive or cross bedded and commonly bioturbated. The *Cruziana* trace-fossil community in sandstones in the Oquirrh Mountain area suggests a shallow-marine environment subject to wave energy. In the same area, ripple-laminated and cross bedded sandstones were interpreted as shallow-marine tractive-current deposits and massive sandstones as the deposits of density-modified grain flows in deeper water. Shallow-water sandstones are regressive deposits and limestones as transgressive deposits. The transition from deposition of limestone to deposition of deep-water clastic rocks during transgressions occurred sometime during the Missourian and Virgilian.

The inferred Missourian and Virgilian part of the Weber Sandstone on the Wyoming shelf crops out along the south flank of the Uinta Mountains. Western exposures (west of the Duchesne River area) consist of thick-bedded, rarely cross bedded, calcareous sandstone and minor limestone and dolomitic limestone. The sandstones are interpreted as mainly shallow marine in origin but may also in part reflect re-deposition of eolian dune sediments deposited during the largest regressions. Accordingly, the contact between eolian and shallow-marine facies slightly south of the Duchesne River area is present. Limestones are the inferred maximum transgression deposits in the western Uinta Mountains area.

East of the Duchesne River area, the inferred Missourian and Virgilian part of the Weber Sandstone is dominated by thick units of massive and cross bedded sandstone, and limestone is uncommon. The thick sets of cross bedded strata are eolian in origin and represent regressive deposition. The Weber eolian dune field was transgressed during maximum high stands and that eolian sands were re-deposited by mass flows to form massive and (or) deformed beds. This hypothesis is strengthened by the presence of marine fossils in similar sandy facies in the upper (Wolfcampian) part of the Weber Sandstone. A few transgressions may have resulted in deposition of limestone. The Weber limestones in the eastern Uinta Mountains area are non-marine and formed in inter-dune ponds. To the south, Weber Sandstone strata inter-finger with fluvial (transgressive) and eolian sand-sheet (regressive) deposits of the Maroon Formation that were derived from the northern flank of the Uncompahgre uplift. This facies change is well documented in the Rangely area of northwestern Colorado.

The Missourian and Virgilian of the northern Paradox basin consists mainly of mixed clastic and carbonate rocks of the Honaker Trail Formation (the upper formation of the Hermosa Group) and the lower part of the "Elephant Canyon Formation". These strata were considered as the lower part of the upper member of the Hermosa Formation. Missourian and Virgilian strata in this area may be as thick as about 600 m.

Coarse alluvial clastic rocks of the upper part of the Hermosa Formation crop out on the northeastern margin of the basin. Limestone is the most common lithology in the Honaker Trail Formation and is inferred to represent transgressive deposition (**Figure 28**).

Figure 28. The Honaker Trail formation limestones are exposed in the canyon walls intermixed with varicolored shale, and siltstone. Location is Goosenecks State Park in Utah.

Honaker Trail clastic rocks in the northern Paradox basin consist of varicolored siltstone and shale and uncommon sandstone in the subsurface which are inferred to be deposited during regressions in sabkha to shallow-marine environments. Eolian sandstones occurred from outcrops of the Honaker Trail Formation (assigned sandstones to the upper part of the Hermosa Formation) in the Moab area in the southeastern part of the northern Paradox basin. Thick sandstone beds of possible eolian origin are common in nearby boreholes but pinch out to the north and west. These eolianites are also inferred regressive deposits and may be more abundant in the main part of the Paradox basin to the south.

Relief in the structural zone on the northwestern flank of the Uncompahgre uplift is inferred to have been less than in the Desmoinesian. Missourian and Virgilian strata are inferred to have on lapped the northwest- plunging uplift. This decreased relief probably led to improved circulation through the area of the Oquirrh sag (the eastern Uinta basin). Several depositional facies converge in and north of the Oquirrh sag, an area for which there is essentially no late Paleozoic outcrop or subsurface control.

In the Eagle basin, Missourian and Virgilian strata include part of the non-marine Maroon Formation, and in the center of the basin, the uppermost, non-evaporitic part of the Eagle Valley Evaporite. A Missourian and Virgilian age for the upper Eagle Valley Evaporite (not previously inferred) is based on the following. A 987-m-thick measured stratigraphic section of the upper Eagle Valley Evaporite in the Eagle Valley in the central part of Eagle basin was used to estimate age. This section begins at the top of the highest thick (60 m) gypsum bed in the Eagle Valley Evaporite, lacks unconformities, and is a continuation of a measured stratigraphic section. The section was traced into the upper Desmoinesian Jacque Mountain Limestone Member which forms the uppermost bed in the Minturn Formation from the Minturn area on the east flank of the Eagle basin into the middle of the Eagle basin where it immediately underlies the highest thick gypsum deposits.

In the Minturn area, a Maroon Formation thickness of 1,147 m, using the top of the Jacque Mountain Limestone Member as its base was measured. This Maroon Formation thickness is similar to that of the post-Jacque Mountain section of the upper Eagle Valley Evaporite and the Maroon Formation along the Eagle River (1,047 m), and the two sections are considered correlative. The stratigraphic cross section shows this relationship and illustrates how the upper Eagle Valley Evaporite in the central part of Eagle basin inter-fingers with the lower part of the Maroon Formation on both basin margins. Although lacking diagnostic fossils, the Maroon Formation is considered late Desmoinesian to mid-Wolfcampian in age based on regional considerations. If relatively constant sediment accumulation rates are assumed for the Maroon throughout this period, then the upper Eagle Valley Evaporite (which also lacks diagnostic fossils and, as discussed above, inter-fingers with the lower Maroon) would have a late Desmoinesian to Virgilian age.

The Eagle basin became progressively more non-marine with time throughout the late Paleozoic, a pattern that probably reflects decreasing rates of subsidence and associated increased progradation of clastic rocks from the basin margins. Facies and paleo-current patterns indicate that the axis of Eagle basin was strongly skewed to the northeast during Maroon time, constituting a reversal in basin polarity from the Morrowan and Atokan. The Uncompahgre uplift was the main source of clastic detritus in the Maroon. The Sawatch uplift influenced depositional patterns but probably had low relief.

The Maroon Formation has a maximum thickness of about 1,000 m in the main part of Eagle basin and consists of interbedded fluvial and eolian deposits. Near the basin margins, Maroon fluvial deposits consist of channel architectural elements of conglomeratic sandstone. These proximal channel deposits grade down the paleo-slope into laminated sand elements of very fine to fine grained sandstone. The facies changes reflect decreases in flow strength, depth, and discharge, characteristics of "terminal fan" fluvial systems in arid basins. Maroon Formation eolianites are mainly sand-sheet deposits. These sand-sheet deposits typically comprise about 30 percent of Maroon sections and appear to cyclically alternate with fluvial intervals. Some fluvial deposits were also deposited lateral to the eolian sand sheets in wadi streams.

Silt eroded from the sand-sheet deposits in the Maroon Formation was transported to the downwind basin margin, the northwest flank of the Sawatch uplift (Ruedi Reservoir area) where it was redeposited as a thick section of loess. The inferred non-marine cyclicity in the Maroon Formation was controlled by cyclic climatic changes synchronous with global glacio-eustatic changes. In this interpretation, fluvial deposition dominated in the Maroon during wetter climatic periods (correlating with eustatic high stands), and eolian sand-sheet deposition dominated during drier climatic periods (eustatic low stands) (**Figure 29**).

The upper, non-evaporitic part of the Eagle Valley Evaporite, which crops out along the Missourian-Virgilian axis of the Eagle basin, consists of several gradational facies: (1) plane-bedded or low-angle-bedded, varicolored, fine- to coarse grained sandstone; (2) ripple-laminated siltstone to fine grained sandstone; (3) massive, extensively bioturbated sandstone and siltstone; (4) grayish-black, thinly laminated, calcareous siltstone and mudstone; and (5) grayish-black micritic limestone and lime mudstone. These facies represent a continuum from sabkha and shallow-water (regressive) to offshore marine (transgressive) deposition. As in the late Atokan and Desmoinesian, the Eagle basin was probably completely cut off from normal marine waters during regressions, when the central part of the basin may have resembled a playa lake. During transgressions, marine waters entered the basin from the north. Marine limestone of inferred Missourian and Virgilian age (based on stratigraphic position) is present at Ripple Creek, Miller Creek, and Sylvan Creek. The contact between these limestones and sandy marine environments (the transgressed Weber sand sea) in northwestern Colorado is in the subsurface.

In summary, the Missourian and Virgilian in the Uinta-Piceance region was characterized by decreasing tectonic activity reflected in lower rates of subsidence. In the Eagle basin, fluvial and eolian deposition dominated while evaporite deposition ended. Evaporite deposition similarly ended in the Paradox basin and was replaced by deposition of sabkha and (or) shallow-marine deposits during regressions. Progradation of sands across the Wyoming shelf continued during regressions. Transgressive deposition of limestone was limited to the western part of the Uinta-Piceance region and a small area in the eastern Eagle basin.

Figure 29. The Ruedi Reservoir area, the Maroon Formation clastics overlie the Eagle Evaporites (center). The Maroon Formation is the reddish shading in the hillsides surrounding the exposed white gypsum deposit.

Early Early Permian: Early Wolfcampian

The major basins of the Uinta-Piceance region continued to subside during the early Wolfcampian. As the Emery uplift subsided, however, the Paradox basin became connected to the Callville shelf and its history as a narrow trough ended. Subsidence exceeded sediment supply in the southeastern Oquirrh basin, so that it was both a depositional and topographic basin. Depositional patterns on the Wyoming shelf and in the Eagle basin remained similar. Major eustatic changes in the Wolfcampian with inferred magnitudes of as much as 170 m occurred.

The Pakoon Limestone (50-70 m thick) represents early Wolfcampian deposition on the Callville shelf; the Queantoweap Sandstone (50-90 m thick) represents late Wolfcampian deposition. Few lithologic descriptions are available for the Pakoon in the Uinta-Piceance region.

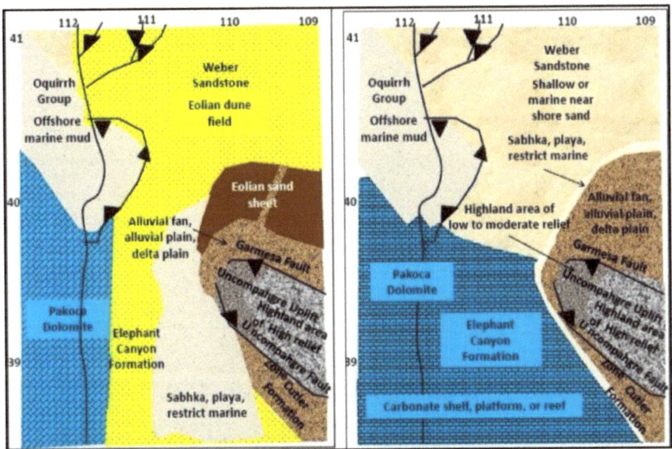

Figure 30. Maps showing the early Wolfcampian paleogeography of the Uinta-Piceance basin region during maximum regression (left) and transgression (right).

It apparently consists of dolomitic limestone and minor sandstone (Cove Fort area outcrops), inferred to represent transgressive and regressive deposition, respectively. Features of the Pakoon carbonate rocks indicate a somewhat restricted, shallow marine environment. Pakoon sandstones are cross bedded. They resemble sandstones in the unconformably underlying Callville Limestone. These sandstones could be shallow marine or possibly eolian in origin. Because Wolfcampian subsidence rates on the Callville shelf were apparently very low (suggested by the small thickness of preserved section), it is also possible that the Callville shelf was an emergent, erosional area during larger regressions (**Figure 30**).

There are apparently no transitional facies between Pakoon and Queantoweap strata in the Cove Fort area and the thick section of Oquirrh basin Wolfcampian strata to the north in the Route 148 southwest area (**Figure 31**). The Pakoon Limestone apparently grades eastward into the upper part of the Elephant Canyon Formation in the northern Paradox basin. These latter strata were considered the Rico Formation and the upper part of the upper member of the Hermosa Formation and "lower Cutler beds". Although debate concerning the stratigraphic nomenclature continues, the name Elephant Canyon Formation is used here for convenience and because of its prevalence in the literature. In the main part of the northern Paradox basin, the Elephant Canyon Formation is as thick as about 460 m and consists of interbedded carbonate rocks, fine-grained clastic rocks, and minor anhydrite in the subsurface.

Figure 31. North of UT 148 on UT 143, exposures of clastics and limestone appear on the west side of the road along the canyon slopes. View is facing northwest. UT 143 is located on the rim of the exposure. Strata are of Wolfcampian age, part of the Oquirrh Basin deposits.

Limestones are commonly fossiliferous and are interpreted as shelf sediments deposited during transgressions. Interbedded clastic rocks are interpreted as sabkha, playa, and shallow-marine regressive deposits.

Sandstones of inferred eolian origin are also present at a few locations in the upper part of the Elephant Canyon Formation in the southern part of the northern Paradox basin and in a narrow belt near the contact with the Cutler Formation on the east flank of the basin. Cycles (about 6-40 m thick) consisting of marine and erg-margin deposits in the upper part of the Elephant Canyon Formation in the central to northern Paradox basin. Transgressive deposits consist of normal-marine fossiliferous shelf limestone and tidal sandstone and are as thick as a few tens of meters. Regressive deposits consist of fluvial and eolian sandstone. The relatively sharp lower contacts of limestone beds indicate rapid eustatic rises and inundation of erg margins. The marine incursions are thought to have entered the Paradox basin from the west. Closely spaced (every 4-15 m) super-surfaces in the eolian and fluvial units may have formed during wet climatic intervals by river flooding. Based on the abundance of the super-surfaces, climatic fluctuations of shorter periodicity than late Paleozoic eustatic fluctuations may have influenced sedimentation in this area.

The Cutler Formation comprises coarse alluvial conglomerate and arkosic sandstone derived from the Uncompahgre uplift. As in the Missourian and Virgilian, narrowing of the width of the band of coarse elastics derived from this highland to the northwest suggests diminished relief on the Uncompahgre in that direction. On lap of the northwestern flank of the Uncompahgre uplift apparently continued in the Wolfcampian.

In the area of the Emery uplift, transitional strata between the Pakoon Limestone to the west and the upper part of the Elephant Canyon Formation to the east consist of dolomite and limestone, thick to thin-bedded sandstone, varicolored siltstone and shale, and minor anhydrite in the subsurface. The proportion of sandstone is variable but probably is significantly greater than in the upper part of the Elephant Canyon Formation. Most likely the sands were not transported westward across the Paradox basin from the Uncompahgre uplift but instead had northern sources in sand-rich Weber environments of the Wyoming shelf or southern sources in the eolian sands of the Cedar Mesa Sandstone. The sedimentology of these Elephant Canyon sands is unknown. They were deposited by eolian processes during larger regressions and graded westward into shallow-marine environments. The Cedar Mesa Sandstone inter-fingers with the upper part of the Elephant Canyon Formation to the south in the main part of the Paradox basin and overlies the upper Elephant Canyon Formation in the northern Paradox basin.

Lower Wolfcampian strata in the southeastern Oquirrh basin include the Curry Peak and Freeman Peak Formations and undifferentiated strata of the Oquirrh Group. These Wolfcampian strata are as thick as 1,500 m thick and are unconformably overlain (in sequence) by the Diamond Creek Sandstone of Wolfcampian age and the Park City or Phosphoria Formations of late Early (Leonardian to Wordian) Permian age.

Lower Wolfcampian rocks are mainly massive, cross bedded, graded, and ripple-laminated sandstone with minor inter-beds of conglomerate and limestone. Conglomerate crops out in a northwest trending band along the southwest margin of the basin and contains limestone clasts older than host strata. The sandstones and conglomerates were interpreted as the deposits of sediment gravity flows, including turbidity currents, density-modified grain flows, and debris flows. Trace-fossil communities in the sandstones suggest that deposition occurred below wave base at bathyal depths. A slope or basin-floor depositional environment occurred at depths of a few hundred meters or more. These depths exceeded the magnitude of inferred sea level changes and thus provided a rationale for similar depositional processes and environments during both eustatic high stands and low stands. Low stand deposits are probably more conglomeratic, containing clasts derived from periodically exposed platforms and carried basin ward through erosional channels.

The location of the transition from shallow- to deep-water environments on the northeastern basin flank is based on the maps produced west of the Wasatch fault and on the facies descriptions in the area of the Charleston-Nebo thrust sheet.

On the southwestern flank of the basin, this transition must lie northeast of the Lower Permian limestones of the Oquirrh Group in the Route 148 southwest area.

The upper part of the Weber Sandstone consists almost entirely of sandstone and comprises the early Wolfcampian of the Wyoming shelf depositional province. It is unconformably overlain by the Park City Formation. Sandstones of the Weber are predominantly cross bedded in the eastern Uinta Mountains (east of the Duchesne River area) and predominantly massive in the western Uinta Mountains. These strata were probably deposited in eolian (maximum regressions) and shallow marine (maximum transgressions) environments, and the relative proportions of the two bedding types reflect the dominant depositional process. Shallow-marine conditions probably dominated in the western Uinta Mountains. Eolian dunes probably formed only during a few of the larger regressions and were rapidly reworked in subsequent transgressions (**Figure 32**).

Figure 32. Exposures of upper Weber Sandstone represent the Wyoming Shelf of Wolfcampian age. Source: Flickr posted on the internet.

Similarly, the large proportion of cross bedded strata in the eastern Uinta Mountains suggests that eolian conditions dominated and that only the largest transgressions submerged the Weber dune field. The presence of marine fossils, massive beds, contorted stratification and extensive burrowing, all support intermittent marine conditions for the Weber in the eastern Uintas.

The Weber Sandstone inter-fingers to the south and southeast with fluvial and eolian sand-sheet deposits of the Maroon Formation in the subsurface which were derived from the northern flank of the Ancestral Uncompahgre uplift.

The upper part of the Maroon Formation comprises the early Wolfcampian of the Eagle basin. These strata are unconformably overlain by the Upper Permian to Lower Triassic State Bridge Formation. Similar to the Missourian and Virgilian lower part of the Maroon, these strata consist of mixed fluvial and eolian sand-sheet deposits in most of the basin. The basin continued to be asymmetric, with its depositional axis skewed strongly to the east. During transgressions, fluvial systems draining the Uncompahgre and Front Range uplifts flowed across the basin, converged along the depositional axis, and flowed north to exit the basin. Rare marine beds from the upper Maroon in the Sylvan, Miller Creek, and Ripple Creek areas showed dolomitic and gypsiferous mudstones from the uppermost Maroon in the subsurface. These strata were probably deposited during the largest transgressions in a sabkha or marginal-marine setting (**Figure 33**).

Figure 33. The Maroon Bells in Colorado exposes the Maroon Formation along its upper slopes. The formation consists of siltstone.

As in the Missourian and Virgilian, eolian sand-sheet deposits were dominant in the Eagle basin during regressions and merged to the north with the Weber dune deposits. Loess continued to be deposited in the Ruedi Reservoir area on the downwind basin margin adjacent to the low-relief Sawatch uplift.

Near the end of Maroon deposition, however, deposition of eolian dune sediments of the Fryingpan Member of the Maroon Formation over the loessite formed a basin-margin dune field. The termination of loess deposition and the initiation of dune deposition were probably forced by a cessation or slowing of subsidence in the Eagle basin, the source area for the basin-margin eolianites.

Rather than being rapidly buried beneath the Maroon alluvial-eolian plain, sands in the source area were exposed at the surface and made susceptible to eolian erosion and transport for longer time intervals. Consistent with this interpretation, thick beds of coarse to granular sand interpreted as deflation lags are present at the top of the Maroon Formation in many places.

In summary, depositional patterns initiated in the Missourian and Virgilian in the Eagle basin and on the Wyoming shelf continued into the early Wolfcampian. The Emery uplift became fully or mostly submerged and the Paradox basin ceased to be a discrete geomorphic element. Deep water clastic deposition began in the Oquirrh basin (or continued from the latter part of the Missourian and Virgilian). Transgressive limestone deposition was limited to the southeastern part of the Uinta-Piceance basin region.

Discussion

The Oquirrh basin experienced the greatest rates and magnitude of subsidence, the Eagle and northern Paradox basins experienced intermediate and similar rates, and the Wyoming shelf experienced the lowest rates and magnitude of subsidence. All four sedimentary provinces experienced their lowest rates of subsidence in the Early Pennsylvanian, their highest rates in the Middle Pennsylvanian, and intermediate rates in the Late Pennsylvanian and Early Permian. The decrease in subsidence rates from the Middle to Late Pennsylvanian in the Oquirrh basin is minor in contrast to decreases in the other three depositional provinces.

Subsidence in the northern Paradox basin is inferred to have been mainly controlled by flexural loading driven by westward over thrusting of the Uncompahgre uplift. As stated earlier, the basin axis was skewed strongly to the east adjacent to the Uncompahgre uplift consistent with this inference.

Subsidence in the eastern Eagle basin closely resembles that in the northern Paradox basin, supporting the contention that the Gore fault (for which the sense of late Paleozoic offset is uncertain) on the eastern margin of Eagle basin was also a reverse fault. Alternatively, two different yet synchronous tectonic processes (flexural loading in the northern Paradox basin, oblique or normal rifting in the Eagle basin) would need to be invoked to produce their almost identical subsidence histories.

Evaporite deposition in both the northern Paradox basin and the Eagle basin probably was controlled by subsidence rate, drainage restriction, arid climate, and fluctuating eustasy. Evaporite deposition in each basin occurred only during the Middle Pennsylvanian, the time of most rapid subsidence. Rapid subsidence commonly results in trapping of clastic sediment adjacent to basin margins causing basin interiors to become isolated from supplies of clastic sediment (or sediment starved), a partial requirement for evaporite deposition.

Drainage restriction took different forms in the two basins, with an inferred tectonic barrier at the northern end of the Paradox basin and a combined tectonic (slowly subsiding) and sedimentary (the positive relief on the Morgan Formation dune field) barrier at the northern end of the Eagle basin. Regular eustatic fluctuations led to submergence of these barriers and replenishment of the water column. Re-emergence of barriers during eustatic regressions led to isolation of water bodies and climate-induced evaporite deposition. All of the above conditions were required for evaporite deposition; evaporite deposition ceased in the Late Pennsylvanian when subsidence rates diminished, and clastic sediments began to prograde into the basin interiors.

The slowly subsiding Wyoming shelf served as a bypass zone across which sediment was transported southwestward to the rapidly subsiding Oquirrh basin. The much larger amount of total subsidence in the Oquirrh basin (6-7 times greater than on the Wyoming shelf) suggests that significant structures (faults or ramps) were present between the two depositional provinces. This inferred zone of structural weakness is not presently recognizable. It was presumably reactivated during the Cretaceous Sevier and early Tertiary Laramide orogeny or has been buried by younger sediments. Despite the contrasts in subsidence history, shallow-marine deposition in both the Wyoming shelf and the Oquirrh basin through most of the Pennsylvanian indicates that sediment supply kept up with subsidence in the Oquirrh basin. Deepening of the Oquirrh basin in the latest Pennsylvania and the Early Permian was apparently not accompanied by a significant increase in subsidence rates. Diminishing sediment supply may therefore have led to the formation of deep-water depositional environments.

The Oquirrh basin is distinguished from the three other depositional provinces on the basis of its higher rates and magnitude of subsidence and its continuing high subsidence rates through the Late Pennsylvanian and the Early Permian. The entire Uinta-Piceance basin region was located east of the isopleth for Mesozoic and Cenozoic igneous rocks, thus its basement rocks are similar.

The Oquirrh basin, however, does overlie the hinge zone of the Late Proterozoic to mid-Paleozoic miogeocline, and weakened crust in this zone might in part be responsible for the anomalously high subsidence in the Oquirrh basin. More importantly, the tectonic controls on subsidence in the Oquirrh basin were probably different from those in the Eagle and Paradox basins.

Both the timing and geometry of subsidence suggest that late Paleozoic deformation in the Uinta-Piceance region of the Ancestral Rocky Mountains resulted from interactions along both the western and southeastern continental margins. As stated earlier, initial subsidence of the Oquirrh basin and the Wyoming shelf as discrete basinal elements began in the Late Mississippian.

This subsidence predates the onset of significant deformation in the more proximal parts of the foreland province associated with the Marathon-Ouachita orogenic belt (the collisional mountain belt of the southeastern convergent margin) by 10-20 million years and predates initial subsidence in the Eagle and Paradox basins by 25-30 million years.

Assuming that the effects of the continent-continent collision to the southeast propagated inland with time as the collision broadened or even that deformation was initiated synchronously across the entire foreland, the early phases of subsidence in the Oquirrh basin and Wyoming shelf are too old to be the result of the collision and demand an independent driving force. The inferred extensional or trans-tensional origin of the Oquirrh basin is consistent with the structural style noted to the west, and a western driving force is highly likely. In contrast, the inferred contractional or trans-pressional style of the Paradox basin and possibly the Eagle basin may be more consistent with the foreland deformation associated with a convergent margin.

Can the effects of the western and southeastern driving forces be isolated? Because significant subsidence in the Eagle and Paradox basins did not begin until the Atokan, these basins may never have been significantly affected by the western driving force. The Desmoinesian acceleration of subsidence in the Eagle, Paradox, and Oquirrh basins is roughly correlative and therefore may in each case reflect the growing influence of the southeastern force (the dominant influence in the Eagle and Paradox basins and a supplemental influence in the Oquirrh basin). The Late Pennsylvanian decrease in subsidence in the Eagle and Paradox basins indicates that the influence of the southeastern force diminished; continued rapid Late Pennsylvanian and Early Permian subsidence in the Oquirrh basin supports the continued importance of west-induced extensional deformation in the western Uinta-Piceance region during this interval.

While considering only interactions on the southern continental margin as a driving force for deformation in the Ancestral Rocky Mountains, the high angle between the uplifts and basins of the Ancestral Rocky Mountain orogeny and the Marathon-Ouachita Orogenic belt requires consideration. A rationale for these anomalous trends in the context of northward-imposed paleo-stress, the geometry of structures in at least the western Ancestral Rocky Mountains is more consistent with paleo-stress imposed by trans-tensional deformation along a western continental margin. Regional trans-tension in the Ancestral Rocky Mountain province is supported by the mix of compressional and extensional structural styles.

Future work on Pennsylvanian and Permian deformation and subsidence in the western United States (including locating and describing the structures bounding the Oquirrh basin), along with better documentation of modern analogues, will further clarify the nature of intra-plate deformation in the Ancestral Rocky Mountains.

Conclusions

Expansion and contraction of late Paleozoic Gondwana ice sheets resulted in repetitive global eustatic and climatic fluctuations. These eustatic and climatic fluctuations, along with tectonic activity and variations in sediment supply, were major controls on the development of transgressive-regressive depositional sequences in the Uinta-Piceance basin region of northwestern Colorado and northeastern Utah.

Major changes in depositional patterns in the Uinta-Piceance region associated with the repeated transgressions and regressions are shown on four sets of paleogeographic maps representing key time periods in the late Paleozoic. The contrasts between maps representing maximum transgressions and regressions are substantial and indicate the necessity of this approach for adequately representing late Paleozoic paleogeography.

The Uinta-Piceance basin region includes four major sedimentary provinces: the Eagle basin, the northern Paradox basin, the southern Wyoming shelf, and the southeastern Oquirrh basin. The Oquirrh basin experienced the greatest rates and magnitude of subsidence. The Eagle and northern Paradox basins experienced intermediate and very similar rates, and the Wyoming shelf experienced the lowest rates and magnitude of subsidence. All four sedimentary provinces experienced their lowest rates of subsidence in the Early Pennsylvanian, their highest rates in the Middle Pennsylvanian, and intermediate rates in the Late Pennsylvanian and Early Permian. The timing, magnitude, and geometry of subsidence in the Uinta-Piceance region suggests that regional trans-tension was driven by the overlapping influences of a more distant, convergent continental margin to the southeast and a more proximal, transform-fault (?) continental margin to the west. The effects of this complex intra-plate deformation are reflected in the considerable variations within transgressive-regressive depositional sequences.

Morrowan and lower Atokan strata throughout most of the Uinta-Piceance basin region consist of alternating fine grained clastic rocks (regressive deposits) and more abundant limestone (transgressive deposits). Significant Morrowan and early Atokan tectonic uplifts include the Ancestral Front Range and Sawatch uplifts. Intra-basinal faulting characterized the Eagle basin and locally influenced deltaic and prodeltaic deposition.

The late Atokan to Desmoinesian history of the Uinta-Piceance region was characterized by increased tectonic activity which is reflected in higher rates of basin subsidence, continued unroofing of the Front Range and Sawatch uplifts, and initial uplift of the Uncompahgre uplift. The combined effects of tectonism and eustasy led to restricted circulation and evaporite deposition in the Eagle and Paradox basins.

Regressive deposition was also characterized by significant progradation of eolian sands across the Wyoming shelf en route to the Oquirrh basin. Limestone deposition dominated during transgressions.

The Missourian to Virgilian history of the Uinta-Piceance basin region was characterized by decreasing tectonic activity and rates of subsidence. Progradation of clastic sediments into the central parts of the Eagle and northern Paradox basins resulted in cessation of evaporite deposition. During regressions, fluvial and eolian deposition dominated in the Eagle basin whereas sabkha and (or) shallow-marine deposition dominated in the northern Paradox basin. The Wyoming shelf continued to serve as a conveyor belt of clastic sediment to the Oquirrh basin, which was characterized by a transition from shallow- to deep-water deposition. Transgressive deposition of limestone was limited to the western part of the Uinta-Piceance basin region and a small area in the eastern Eagle basin.

Depositional patterns initiated in the Missourian and Virgilian in the Eagle basin and on the Wyoming shelf continued into the early Wolfcampian. The Emery uplift was fully or mostly submerged, and the Paradox basin ceased to be a discrete geomorphic element. Deep-water clastic deposition continued in the Oquirrh basin. Transgressive limestone deposition was limited to the southwestern part of the Uinta-Piceance region.

Chapter 5. The Uinta Basin

Introduction

Regional unconformities and thick sedimentary basins in the central Rocky Mountains, Colorado Plateau, and northeastern part of the Great Basin preserve an impressive record of late Paleozoic ("Ancestral Rockies") orogenesis. Many regional structures of this age have been obscured by younger tectonism, basin development, and volcanism. With few exceptions, major late Paleozoic faults at the margins of Precambrian exposures have been reactivated or overprinted by Laramide-age (Late Cretaceous to Eocene) or younger faults. In order to understand late Paleozoic faulting, it is necessary to separate younger movement from Paleozoic movement or to identify un-reactivated Paleozoic structures. Identification of late Paleozoic faults that have not been reactivated (or have undergone only minor reactivation) is best accomplished using subsurface (well and seismic) data. In this chapter, seismic reflection data from the southern part of the Uinta basin, near Price, Utah, are used to illustrate the style of late Paleozoic deformation northwest of the Ancestral Uncompahgre uplift. The seismic profile transected to the northwest direction from Sunnyside through Price, and Helper to the Clear Creek graben ending at the Fish Creek graben in Utah.

Regional Late Paleozoic Stratigraphy; Correlation with Reflection Data

A seismic reflection transect line for the upper Paleozoic rocks of the study area are shown in **Figure 34**. These strata are penetrated by deep wells at several locations, and key reflections from this interval can be traced on seismic reflection lines that can be tied to these wells. Identifications of key stratigraphic horizons were calibrated using sonic logs and velocity surveys in the wells.

The deepest Paleozoic reflector generally recognized in the area is the top of the Mississippian carbonate rocks correlative with the Redwall and Leadville Limestones. The Madison Limestone in the area north of the Emery uplift, and the name Deseret Limestone is commonly used in industry logs of well and seismic data in the vicinity of Price. These rocks will be called Redwall Limestone. On the seismic lines examined in this study, the Redwall reflection lines are a doublet that cannot be traced entirely across any of the seismic lines, probably because of signal penetration problems. The relatively unreflective interval above the Redwall represents the Humbug Formation and the Doughnut Formation (commonly referred to as the Manning Canyon Shale on well logs in the study area).

Figure 34. Seismic reflection transect line from Sunnyside through Price and Helper to the Clear Creek and Fish Creek grabens.

A well-defined doublet at the base of an overlying set of banded reflections marks the top of the Doughnut. The banded reflection sequence corresponds to mixed clastic and carbonate rocks of the Pennsylvanian and Permian "undifferentiated Weber and Morgan Formations". A prominent reflection from the Kaibab Limestone (of Leonardian and locally Guadalupian age) lies above the Weber and Morgan reflections. In this study, local thickness variations are documented in the interval between the Doughnut and Kaibab reflections. This interval is occupied mainly by the Weber and Morgan Formations. In many cases, the thickness of the Weber and Morgan abruptly changes across Pennsylvanian or Permian faults. On seismic sections, these faults displace Mississippian reflections but do not displace the Kaibab reflection. They are the primary evidence used to interpret the late Paleozoic structural evolution of the region and are mapped in the subsurface in this study. On some seismic sections, a lower Permian reflection is identified within the Pennsylvanian and Permian section. This reflection is tentatively correlated with the top of an interbedded marine limestone, sandstone, and shale unit of Virgilian and Wolfcampian age (lower Cutler beds includes rocks assigned to Elephant Canyon Formation). Similar to the Kaibab, the reflection is not typically cut by late Paleozoic faults.

Sonic logs and velocity surveys in wells along the seismic lines were used to verify the correlation between reflections and the stratigraphic section. Two-way travel times for the Doughnut to Kaibab interval were measured along these lines, and these travel times were converted to the thicknesses. This conversion was calibrated using several deep wells (Hunt 1-16 State, American Quasar 31-1 Drunkards Wash, Arcadia Telonis no. 1) that penetrate upper Paleozoic rocks near the reflection lines. Because the average seismic velocity for this stratigraphic interval differs in each of these wells and would certainly vary throughout the study area, thickness values are approximate.

Fault geometry and sense of offset along late Paleozoic faults were determined based on identification of hanging wall and footwall cutoffs of Mississippian reflections. Faults were identified on more than one seismic line in almost every case. Identification of several of the faults on three or more seismic lines permitted detailed observation of along strike variations in offset and fault geometry.

Seismic data that illustrate the structural styles are presented as vibroseis data. Along these seismic lines, the Doughnut to Kaibab interval gradually thickens from southwest to northeast from approximately 1,500 to 3,600 ft (457-1,097 m) and three northeast-directed reverse and thrust faults intersect the lines. The geometry of each of these faults is constrained by hanging-wall and footwall cutoffs of Mississippian reflectors (Redwall and Doughnut). Based on these relationships, three faults dip approximately 30°, 50°, and 10° SE, respectively. Up dip from the first fault, a local increase in dip in the Mesozoic reflectors indicates minor Laramide (?) reactivation of this fault.

Local discordances (truncation or on lap) within the Pennsylvanian and Permian banded zone attest to uplift of individual fault blocks. Near the Arcadia Telonis no. 1 well, inclined reflections in the banded Pennsylvanian and Permian sequence in the hanging wall of the first fault are locally truncated. The reflection appears to be folded above, and perhaps offset by the fault. A dipping reflection in the interval between reflections is truncated and overlain by sub-horizontal reflections. To the north, the reflection continues undeformed over the second and third faults. Just south of fault C, there is a southward-on lapping pattern at a lower stratigraphic level within the banded Pennsylvanian and Permian reflections. Thus, within these two seismic lines, several local unconformities provide evidence for localized erosion and (or) non-deposition at slightly different stratigraphic positions.

Lines near Helper Utah illustrates gradual northwesterly thinning of the Doughnut to Kaibab interval on the up thrown (southeast) side of a minor steeply dipping fourth fault. The fault geometry is not well constrained by the seismic data here, and the dip of Doughnut reflections away from the fault in both directions suggests that an arch is locally developed. Southeast of the fault, a local angular unconformity is defined by the on lapping of Pennsylvania and Permian strata above the gently southeast dipping Doughnut reflector. It is possible that fault-induced arching produced a submarine, or barely emergent, topographic high across which younger layers on lapped.

Arching of reflectors above the fourth fault in the Kaibab through Dakota section demonstrates that post-Paleozoic deformation persisted in this zone and probably included Laramide reactivation of the fault.

A 26-mi (42 km)-long, north-south seismic line in the western part of the study area is inferior to that of the other seismic lines discussed here, especially in an 8-mi (13-km) interval near Schofield Reservoir. The poorer data quality may be related to ray-path complexities produced by young structures. The line that follows the Clear Creek graben, and the Fish Creek graben crosses the seismic line near Schofield Reservoir. The data quality deteriorates significantly in the Fish Creek graben, and the graben itself is not well imaged. The Mesozoic or younger faults may be related to the Clear Creek graben. The seismic line is highly oblique to the strike of this north trending fault zone (**Figure 35**).

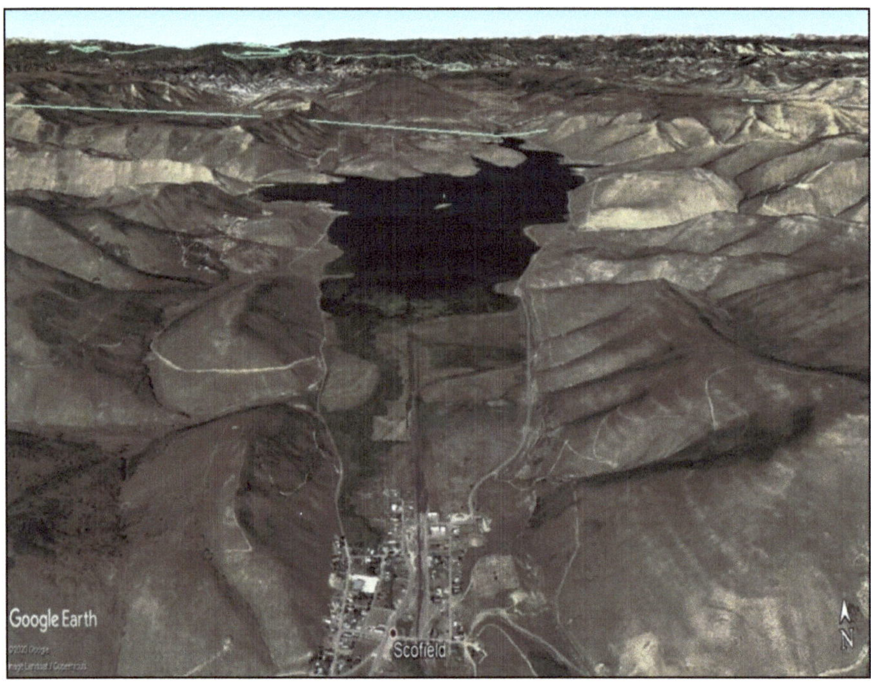

Figure 35. The Clear Creek graben is the location of the northwestern segment of the reflection line transect. The graben itself is faulted by a northeast to southwest trend in the lower left center of the imagery.

Although the Paleozoic section is not well imaged on line, some significant trends are apparent. The southern part of the line is on the north flank of the Emery uplift, as evidenced by the absence of the Doughnut Formation along the southern one-third of this line and in the nearby Arco Hiawatha no. 1 well. North of the fifth fault, the Redwall to Kaibab interval abruptly thickens and discontinuous Doughnut reflections may be present. In the northern part of the line, there are obvious Doughnut reflections.

The abrupt variation in thickness of the late Paleozoic section across the fifth fault indicates that the fault was active during late Paleozoic time. Other faults on this seismic line may have undergone Paleozoic movement, but all of these faults also underwent younger movement, as shown by the offset of the Kaibab and younger reflectors. Thus, it is difficult to obtain a clear picture of late Paleozoic movement for these faults.

Structural Patterns and Thickness Variations

In most of the study area, the thickness of the Doughnut to Kaibab interval varies from about 1,500 to about 4,500 ft (457-1,372 m); the variation is principally controlled by local syn-depositional faulting. The interval is thinnest (1) west and southwest of Helper, corresponding to the north flank of the Emery uplift, and (2) around Sunnyside, at the northwest end of the Ancestral Uncompahgre uplift. West of Price, the principal Paleozoic faults are northwest-striking, northeast-directed reverse faults that step off of the northern and northeastern margins of the Emery uplift and have as much as 5,000 ft (1,525 m) of throw. Across one of these faults, the preserved thickness of the Doughnut to Kaibab interval varies from zero (where the entire interval has been locally eroded) to more than 5,000 ft (1,525 m). Steep northwest-striking normal faults are also present.

North of Helper, a thickened late Paleozoic section in an east-west-striking sub-basin occupies a footwall position with respect to two bounding reverse faults that have opposing displacement directions. The more northerly of the two reverse faults merges to the southeast with a northerly striking, west-directed thrust system that controlled late Paleozoic deformation east and northeast of Price. The structural style in this thrust system resembles that of a foreland thrust belt, in contrast to the block-faulting style that is evident to the west. Imbricate thrust structures were observed on a seismic line about 6 mi (10 km) east of Price. A few back thrusts as well as high-angle reverse and normal faults are also present. The reflection patterns indicate that many of the late Paleozoic faults in this region are essentially growth faults, regardless of geometry (normal, thrust, or reverse). Although local truncation and on lapping relationships within the Morgan and Weber Formations are indicative of local emergence of fault blocks, most of the faulting probably occurred in a relatively shallow submarine setting during Pennsylvanian and Permian deposition and did not result in widespread subaerial exposures.

Mesozoic and Tertiary reactivation and overprinting are most apparent near the Clear Creek and Fish Creek grabens and near Sunnyside. Young faulting occurred in other parts of the study area, but in most cases the late Paleozoic structures have not been greatly modified by younger faulting.

Discussion

Fault patterns in the study area present a complex picture dominated by high-angle (block) faulting in the western part of the study area and by thrust faulting in the eastern part. Reverse and thrust faulting are predominant, but normal faults are locally important. A distinctive style of faulting is associated with each of the two Ancestral Rockies uplifts that bound the study area. Northwest-striking, steeply southwest-dipping reverse faults bound the Emery uplift on its northern and northeastern sides, whereas a north-striking west-directed local thrust system bounds the Ancestral Uncompahgre uplift at its northwest end. The other faults may comprise an interference pattern between these two principal zones, as will be discussed later in this section.

The northwesterly strike of reverse faults that step off the north end of the Emery uplift is almost parallel to the strike of the Uncompahgre fault, known to be a major late Paleozoic reverse fault. The faults are also parallel to the N. 60° W.-striking faults within the Emery uplift. Although the Uncompahgre fault dips northeast and the faults north of the Emery uplift dip southwest, the orientation of all of these faults is compatible with regional northeast-southwest shortening. West-directed thrusting at the northwest end of the Uncompahgre uplift requires that the Uncompahgre block moved west with respect to the late Paleozoic sedimentary basin in the study area. Such movement agrees with the conclusion that the Uncompahgre fault experienced significant strike-slip motion in addition to its well documented reverse slip. Several miles of left-lateral motion on the Uncompahgre fault could have been transferred to the west-directed thrust system, in which shortening occurred in front of the west driven Uncompahgre uplift.

The curved fault pattern north and northwest of Helper is part of a complex zone of accommodation between the Uncompahgre and Emery uplifts. The two uplifts are bounded by en echelon reverse fault zones that require either a strike-slip zone or a complex curved fault zone between the two uplifts. Northwest-southeast motion of the Uncompahgre or Emery blocks would further complicate the structure in the intervening region. Along the curved faults north and northwest of Helper, high-angle reverse displacement (and local low-angle thrusting) occurred on east to northwesterly striking segments. Individual reverse faults pass into low-angle thrusts at their southeast ends; to the south, this fault geometry is replaced by the north-south thrust system described above. The curved faults in this accommodation zone appear to have been flattened and rotated toward the orientation of the north-south thrust system. Such rotation is a further expression of shortening associated with northwesterly movement of the Ancestral Uncompahgre uplift.

The fault pattern in and near the study area is consistent with a general northeast-southwest orientation for regional compression. Slight variations in the regional stress field during Pennsylvanian and Permian time could have produced alternating reverse and strike-slip motion on major pre-existing zones of weakness that trend about N. 60° W.

For example, regional compression oriented between N. 45° E. and N. 15° E. could produce reverse displacement on the Uncompahgre fault and the northwesterly trending faults north of the Emery uplift; easterly excursions of regional compression direction would produce sinistral strike slip on the same faults. The reflection data provide information only on dip-slip motion; there may have been a considerable strike-slip component of motion on some of the northwesterly striking subsurface faults in the study area.

Normal faults are also present, both parallel and orthogonal to the dominantly northwest trends of the reverse faults. Northeasterly striking normal faults are compatible with the northeast-southwest regional compression direction proposed above. The northwesterly striking normal faults may have developed in response to "post-Ancestral Rockies" (and pre-Kaibab) relaxation of the northeast-southwest compression. It is also possible that movement on some of these faults included a strike-slip component. It is difficult to closely constrain the timing of movement on the late Paleozoic faults discussed here. The faults displace reflectors correlated with the Meramecian and Chesterian Doughnut Formation and do not offset reflectors correlated with the Leonardian Kaibab Limestone. In examples cited above, local unconformities (surfaces of truncation and on lap) were identified within the banded reflection zone in the Weber and Morgan Formations.

These are surfaces of erosion and (or) non-deposition above fault blocks and are overlain by reflectors that are not affected by the late Paleozoic faulting. The reflection is stratigraphically higher than almost all of these local unconformities and postdates most of the faults. If the reflector marks the top of the Virgilian and Wolfcampian interbedded carbonate and clastic sequence, as suggested above, almost all of the late Paleozoic faulting would have to be pre-Wolfcampian. Relationships near the first fault are the only examples of slightly younger Permian deformation cited here. Thus, the late Paleozoic faulting in the study area was certainly post-Chesterian and pre-Leonardian and probably pre-Wolfcampian in most cases. If, as suggested above, the faults are genetically tied to the displacement on the Uncompahgre fault, they were most active in Desmoinesian time because Desmoinesian sedimentation patterns strongly suggest that the Uncompahgre uplift was rising most rapidly at this time.

The Ancestral Rockies orogenesis were interpreted as the remote foreland expression of the Ouachita-Marathon orogeny, which was produced by collision between South America-Africa and a peninsular projection of the North American craton. The Ancestral Rockies were formed along the Wichita mega-shear in response to the collision of eastern North America with Africa (Alleghanian orogeny). These two models have different kinematic implications for the structures in and near the study area.

A right-lateral component of displacement on northwesterly striking faults would be consistent with the first model interpretation whereas a left-lateral component of displacement would be consistent with the second model interpretation. Because the study area is small, the buried fault patterns cannot rule out either model; however, the fault patterns are most compatible with the second model which requires left lateral shear along northwesterly trending faults and east- to northeast-directed compression. The relevant observations in the study area include the reverse component of movement on northwesterly striking faults and the local north-south thrust belt that provides evidence for westward motion of the Uncompahgre block relative to the Paleozoic sedimentary rocks in the Price area. This study produced no evidence, however, that the Uncompahgre fault and the study area are part of a mega-shear along which 72-93 mi (120-150 km) of left-lateral displacement occurred. Instead, our analysis implies that strike-slip displacement on the Uncompahgre fault was probably less than 6 mi (10 km).

Chapter 6. Facies Architecture of the Paradox Basin

The structure, stratigraphy, and paleontology of the Paradox Basin fall into three conformable lithostratigraphic units which compose the basin fill. They consist of carbonate and evaporite facies of the Paradox Formation, the mixed siliciclastic and carbonate Honaker Trail Formation, and the siliciclastic Cutler Group of the undivided Cutler Formation in the proximal basin. The lower portion of the proximal Cutler combines with the evaporite and carbonate facies of the Paradox Formation to create a tripartite Desmoinesian system between 310 and 305 mya during deposition of which the majority of basin subsidence and sediment accumulation occurred. Lithofacies and measured stratigraphic sections of these and overlying facies associations from nine localities across the basin combine with the existing literature to provide the chronostratigraphic interpretation of facies architecture.

Tripartite Desmoinesian Basin Fill

Proximal Cutler Formation

Proximal to the structural front of the Uncompahgre uplift, the Cutler Formation is composed of a thick (up to 5 km) succession of boulder to pebble, arkosic conglomerate and subordinate trough cross stratified, very coarse grained sandstone. This course facies thins and fines gradually towards the southwest away from the structural front of the Uncompahgre uplift. The lower parts of the Cutler Formation inter-fingers with the evaporite facies of the Desmoinesian Paradox Formation (310 to 305 mya) and its upper part grades laterally into the Missourian-Virgilian (305 to 295 mya) Honaker Trail Formation, and Virgilian to Wolfcampian (295 to 260 mya) Cutler Group of the medial and distal basin (**Figure 36**).

The proximal Cutler Formation is usually mapped as part of the undivided Permian Cutler Formation owing to its red color and lateral association with the Cutler Group. Well log correlation of marker shale units, and lateral inter-tonguing relationships with the Middle and Upper Pennsylvanian units suggest that deposition of this proximal unit was also synchronous with deposition of the Desmoinesian Paradox Formation and the Desmoinesian to Virgilian Honaker Trail Formation (**Figure 37**).

The proximal Cutler Formation near Gateway, Colorado was interpreted as parts of a dry alluvial fan composed of sediment shed from the Uncompahgre uplift in a southwestward direction. In the larger Gateway section, the relative abundance of gravelly sandstone facies and clast supported, cross stratified, and imbricated cobble conglomerates, and the paucity of matrix supported debris flow deposits suggests that the outside localized fan-head trenches the Gateway fan dominated by traction flow processes.

Figure 36. The Cutler Group, a lateral equivalent to the Cutler Formation is exposed in this canyon. Source: AMES Geology posted on the internet.

Figure 37. The Cutler Formation consists of both divided and undivided red beds are of Permian age. Source: Wikiwand posted on the internet.

The prevalence of trough cross stratified, clast supported conglomerate and very coarse sandstone in an 80 km wide southeast trending swath just down dip from the southwestern margin of the Uncompahgre uplift suggests that the proximal Cutler Formation was deposited by a system of coalesced stream dominated alluvial fans and fluvial mega-fans that reworked boulder and cobble gravels shed from the crystalline basement of the Uncompahgre uplift. The protracted 30 my deposition of this coarse, thick unit alongside the Uncompahgre uplift recorded the complete contractional development of the basement involved structure.

The Paradox Formation

Evaporite-shale and biohermal carbonate facies of the medial and distal basin combine with the coarse arkosic proximal facies described above to define the tripartite Desmoinesian system that recorded the most active phase of subsidence and contractional tectonism in the Paradox Basin. In the medial basin, the Paradox Formation is a thick 3 km mega-sequence of 29 shale-dolomite-evaporite cyclothems. Although rarely exposed in outcrop, these cycles are identified in well logs from boreholes drilled across the basin (**Figure 38**).

Figure 38. The Paradox Basin exposes the Paradox Formation in multiple layered sedimentary rocks consisting of evaporites, shale, carbonates, and arkoses of proximal, distal, and medial basin facies. Source: Southwest Desert Lover posted on the internet.

Bounded by disconformities, these glacio-eustatically driven cycles were interpreted as subaqueous restricted marine deposits. Differential loading by prograding fans of the proximal basin instigated halo-kinetic rise of the Paradox Formation evaporite soon after their deposition. This process developed salt cored anticlines and accompanied growth structures in the upper Paleozoic and lower Mesozoic section of the Moab Utah area. In many parts of the medial basin, halo-kinesis obscured the original depositional thickness of the evaporite facies, although autochthonous sections of 2.2 km or more have been identified.

Contemporaneous with evaporite deposition, a 350 meter thick succession of shelf carbonates and interbedded sapropelic shales developed on the distal southwestern basin margins. Representative depositional cycles in the distal basin generally began with interbedded, laminated, sapropelic marls and carbonate mudstones and wackestones which coarsen and thicken upward into massive, fossiliferous and chert rich units ranging from wackestone to grainstone in texture.

These phylloid algal foraminiferal, oolitic and bioclastic units often have low amplitude, long wavelength biohermal forms and are frequently capped by cross stratified, oxidized, and/or brecciated beds. Locally, lenticular calcite cemented quartzose sandstones are intercalated with these massive carbonate units (**Figure 39**).

Figure 39. Algal mounds consisting of phylloidal algal mats are present in the Paradox Formation limestone. Source: Semantic Scholar posted on the internet.

Fusulinid and bryozoan faunas preserved throughout the distal carbonate facies ascribe a Desmoinesian 310 to 305 mya age for the Paradox carbonates. The contiguous sapropelic mudstones allow for correlations between distal carbonate and medial evaporite cyclothems. These correlations when coupled with the inter-bedding of coarse arkosic and evaporite units in the proximal-medial basin suggest that the tripartite Desmoinesian system records the onset and maximum development of Paradox Basin subsidence. Variations in Desmoinesian interval thicknesses from 4 km in the proximal basin to 350 meters in the distal basin support interpretations of long wavelength differential subsidence during this time.

Honaker Trail Formation

Following deposition of the Paradox Formation, remaining accommodation space in the distal 150 km of the basin was filled by the Missourian-Virgilian Honaker Trail Formation. Distinguished in outcrop from the distal carbonate facies of the Paradox Formation by its less massively bedded strata, higher siliciclastic content, and abundance of primary physical sedimentary structures, the Honaker Trail Formation is a southwest trending carbonate dominated unit that covered the evaporite shale and biohermal carbonate facies of the Desmoinesian system, thereby preserving the underlying depositional topography (**Figure 40**).

Figure 40. The Honaker Trail Formation forms incised meanders in Gooseneck State Park, Utah. Source: Trial Run Project posted on the internet.

The Honaker Trail Formation is generally composed of successions of gray bioclastic packstones, coated grainstones and calcarenites interbedded with slope forming successions of red, green, and purple siltstone and sandstone with green marl and black, sapropelic calcareous mudstones. Calcarenite and coated grainstone beds often have broadly lenticular geometries and are abundantly cross stratified. Massive carbonate packstones are locally rich in crinoid, brachiopod, fusulinid and bryozoan faunas, as well as red, green, gray, and white chert nodules. Boundstone textures are rare in comparison to the Paradox Formation.

In the upper Honaker Trail Formation (i.e. the Rico Formation and/or Elephant Canyon Formation, and the Lower Cutler beds), typical carbonate units are increasingly intercalated with thick, lenticularly bedded, cross stratified quartzose and mottled red sandstones and siltstones.

The abundance of cross stratified calcarenites, lenticular bedding geometries, and disaggregated and coated bioclasts suggest that the Honaker Trail Formation was deposited by inter-tonguing carbonate shoals and coastal channels on a broad shelf offshore of the terrestrial fan systems that continued to dominate deposition in the proximal basin during Late Pennsylvanian time. The Honaker Trail formation records the transition from a localized Middle to Late Pennsylvanian marine and restricted marine depo-system to the regionally expansive, terrestrial Permian system recorded by the overlying Cutler Group.

Cutler Group

Outside of the proximal medial portion of the Paradox Basin where the undivided Cutler Formation occupies a 5 km thick portion of the stratigraphic section, the Cutler Group is a 530 meter thick complex mosaic of inter-fingering buff, orange, red, and maroon arkosic and quartzose sandstones and siltstones of Permian age, belonging to members and facies of these primarily fluvial and eolian strata.

The Wolfcampian Culter Group of the medial and distal basin is significantly thinner than Desmoinesian-Wolfcampian rocks of the proximal Cutler Formation (**Figure 41**). An interpretation of the structural damming of arkosic sediments by halo-kinetic deformation of the Paradox evaporites in the medial basin was completed through review of the isopachous map patterns. It is unlikely that trapping of coarse material along the structural front of the Uncompahgre uplift by the rapid subsidence rates of the proximal fore deep also contributed to restricting the expansion of the Cutler Group until after the available accommodation space in the proximal basin was filled. Along with time equivalent parts of the proximal Cutler Formation, the Cutler Group records the bypassing of the proximal basin by arkosic red bed sediments shed from the Uncompahgre uplift and deposited by an elaborate Wolfcampian fluvio-eolian depositional system. This basin wide deposition was fuelled by untapped sediment supply coffers in the still high Uncompahgre and precluded further intra-basinal carbonate sedimentation.

Moreover, its conclusion marked the end of sediment accumulation related strictly to localized Pennsylvanian-Permian Paradox Basin subsidence.

cres***Figure 41. The Cutler Formation consists of divided and undivided units within the formational unit. Source: Geology Happens blogger posted on the internet.***

The Paradox Basin as an Intra-foreland Flexural Basin

Foreland basins are elongate regions of potential sediment accumulation that form in response to geodynamic process related to the development of a local orogenic belt. Crustal flexure is modeled as the response to the development of a topographic load on an elastic plate. The clastic response of the foot wall plate to this loading creates a depressed region called the fore deep depo-zone closest to the load, as well as a subdued positive response called a peripheral bulge, or fore-bulge at a distance from the load determined by the rigidity of the flexed crust. Distal from this peripheral bulge, an outer region of minor subsidence and sediment accumulation called the back bulge depo-zone may develop as a result of a secondary flexural depression and aggradation up to or above the fore-bulge.

The architecture of the Paradox Basin fill described resembles other foreland basins. The tripartite Desmoinesian system corresponds in relative time and lithologic architecture with the foreland basin depo-zones. In this interpretation, deposition of alluvial fan, fluvial mega-fan, and braid plain sediments in the proximal fore-deep along the structural front of the Uncompahgre uplift is analogous to similar deposits in ancient and modern foreland basins. The distribution of submarine evaporites in the medial fore-deep of the Paradox Basin and these deposits' correlation with distal shallow water carbonates is best described by a barred basin evaporite model. It is consistent with fore-deep deposition. The synorogenic development of an uplift parallel trend of comparatively thin platform carbonates on the distal basin margin is consistent with the model foreland basin carbonates recognized in the Late Paleozoic foreland Tarim Basin, the Huon Gulf Basin, the Alpine and Appalachian foreland basins. Late stage filling of remaining accommodation space by the mixed shallow water carbonates and red beds of the Honaker Trail Formation and Lower Cutler Beds allowed overtopping of the basin by the terrestrial systems of the overlying Cutler Group. Decreased subsidence rates in the proximal basin and a large remaining sediment supply in the Uncompahgre highlands allowed subsequent widespread dispersal of arkosic Cutler Group sediment throughout the basin. This succession of a tripartite synorogenic depo-system overtopped by post orogenic siliciclastics suggests that the Paradox Basin experienced a late history common to many foreland basins.

The Paradox Basin has been subject to syn-and post-orogenic structural and halo-kinetic deformation, as have many foreland basins. The structural style of deformation and its location on the relatively undeformed Colorado Plateau minimize these complications and provide rare insight into the development of ancient intra-foreland flexural basins.

Shortening estimates from basement involved uplifts are typically low up to 10s of kms in comparison to thin skinned fold thrust belts, even though both of these structural systems have been shown to develop associated with flexural basins. Given the sub-vertical nature of the uplift bounding fault in the northeastern side of the Uncompahgre uplift and the lack of large subsidiary contractional structures outside of the Uncompahgre front, it is likely that the total shortening across the Uncompahgre-Paradox system is not much greater that the 10 km recorded by the basement overhang on the southwestern side of the uplift. This suggests that the 50 km wide Uncompahgre uplift load was largely non-migratory, especially in comparison to the 100s to 1000s of kms load migration values typical of large fold thrust belts. Subsidence curves from individual locations do not display the upward convex patterns characteristic of foreland basins that result from thin skinned, migratory fold thrust belts. The exponential decrease in subsidence magnitude away from the Uncompahgre uplift as recorded in subsidence curves beckons further investigation of the distribution of differential subsidence. The non-migratory nature of the thick skinned Uncompahgre load eliminates the need to conduct complicated basin fill restorations in order to restore the Paradox Basin to earlier states. This simplifies the procedure of flexural modeling described below.

Isopach, seismic, surface geology, and borehole data were compiled to construct a two dimensional profile across the Uncompahgre Paradox system. By choosing the top of the Missourian-Virgilian Honaker Trail Formation as a datum, stratigraphic units deformed by post-Pennsylvanian halo-kinetic and structural uplift were restored to their positions at the end of the Virgilian. The interface between the Paradox Formation and its proximal Cutler Formation siliciclastic equivalents and pre-Desmoinesian strata can then serve as a proxy for the basin's paleo-profile as localized subsidence waned in the Late Virgilian. The flexural equations were applied to match a theoretical flexural profile to the actual basin profile of the Uncompahgre-Paradox system.

Discussion

As a foreland basin, the Paradox Basin presents a challenge to existing models for the ARM and assumptions about the regional tectonic setting of intra-continental flexural basins. Foreland basins develop in convergent tectonic regimes in close proximity between 100 and 1000 km to orogenic belts and have depositional areas approximately perpendicular to regional shortening directions. The NW to SE trends of the Paradox Basin flexural axis, the Uncompahgre uplift, and the trace of the Uncompahgre thrust suggests that the Uncompahgre-Paradox system developed as a result of shortening in a SW to NE direction. The lack of evidence for strike slip offset along the flanks of the Paradox Basin, the Uncompahgre uplift and other ARM basins and uplifts further supports this interpretation. The fact that a number of other ARM uplift basin pairs have orientations, structural styles, and flexural signals similar to the Uncompahgre-Paradox system calls into question models that ascribe development of the ARM study to NW directed convergence transmitted from the southern North American margin.

Whereas structural and geodynamic evidence for crustal shortening in the ARM is strong, transpression cannot be entirely ruled out. It is possible that widespread intra-continental transcurrent deformation could develop local contractional structures large enough to produce large flexural basins. However, given the widespread distribution of ARM uplifts and basins that show SW to NE and W to E shortening, models that better provide for those orientations of shortening should be carefully considered at least until unambiguous evidence for through going strike slip offset is documented across the greater ARM region.

Two other orogenic systems have been compared with the ARM. The Late Cretaceous to Eocene Laramide orogeny of the western interior USA, and the intra-continental deformation associated with the Cenozoic collision of India and Asia. Both of these orogenic systems display widespread basement involved deformation inboard from their respective active margins, and were used as analogs for the ARM. The driving mechanisms in neither of these orogens resemble tectonic processes recognized along the margins of Late Paleozoic North America.

Laramide structures are similar in size between 40 and 80 km wide, and 100s of km long. Structural style of basement involved intra-continental thrusts to ARM uplifts, in many cases Laramide fault slip actually reactivated ARM uplifts. The large loads and small magnitudes of shortening of less than 10 km associated with the Laramide uplifts produced adjacent flexural basins with non-migratory foreland basin depo-zones. In many cases they are similar in size and architecture to those of the Paradox Basin. Most of the Laramide uplifts and basins are oriented perpendicular to the E/NE convergence direction of the shallowly dipping subducted plate that drove their development. ARM uplifts and basins are elongated parallel to the expected shortening direction if indeed shortening was caused by stresses transmitted from the Ouachita-Marathon thrust belt. Pennsylvanian-Permian NE directed flat slab subduction along the southwestern North American margin induced ARM shortening and uplift. Evidence for such a subduction system in the Late Paleozoic is sparse.

The widespread intra-continental deformation in central Asia associated with Himalayan orogenesis also beckons comparison with the ARM. The NW trending structures of the ARM resulted from lateral extrusion of crustal blocks as a Gondwanan promontory penetrated the southern margin of North America similar to the escape tectonic regime recognized in modern central Asia. The central Asian analogue is not appropriate for the ARM for at least three reasons. First, hundreds of kms of strike slip offset have been well documented along many uplift and basin bounding faults in central Asia which is not the case in the ARM. Secondly, Cenozoic deformation in and around the Tibetan Plateau developed on the over-riding Asian plate. The ARM developed on the subducting plate. Thirdly, deformation in central Asia produced a regional, high elevated plateau known as the Tibetan Plateau, topographically and structurally distinct from the isolated basin and range paleo-topography of the ARM. A better analogy for the ARM would be the locally high topography generated by basement involved thrusting in northern and central India known as the Bundelkhand, Aravalli Ranges, and the Sillong Plateau.

Conclusions

The Paradox Basin is an intra-continental flexural basin that developed under the load of the thrust bounded ARM Uncompahgre uplift during Middle Pennsylvanian through Early Permian time. The basin subsided rapidly during the Desmoinesian about 310 to 305 mya when uplift of the crystalline Uncompahgre block developed 2 to 5 km of accommodation space in the proximal basin and one tenth of that on the distal margins. This large topographic load shed coarse granitic and arkosic sediment into the proximal basin, spurring the rise of a peripheral fore-bulge. This induced deposition of thick evaporite-shale successions in the medial basin and growth of carbonate bioherms on the distal basin margins. The transitional Honaker Trail Formation filled remaining accommodation space in Missourian through Virgilian time between 305 and 295 mya.

The large foreland basin system was then overtopped in the Wolfcampian between 295 and 260 mya by the complex terrestrial depo-systems of the Cutler Group which were no longer trapped by rapid subsidence in the proximal basin as they were during the Middle Pennsylvanian. Loading of the Paradox Basin was accomplished by thrust displacement along parallel, oppositely dipping faults on either side of the 50 km wide NW to SE trending Uncompahgre uplift. The moderately NE dipping Uncompahgre fault achieved about 10 km of shortening along the southwestern margin of the uplift while faulting on the northeastern margin accommodated only minor shortening but equally great structural relief of about 5 km along a steeply dipping top to the northeast fault system. The resulting flexural wavelength suggests that at least one of these faults completely penetrates the elastic crust. Although transpression cannot be entirely ruled out, all available data suggests that the amount of strike slip displacement along these faults is minimal, and that deformation is dominantly represented by NE to SW shortening. The Paradox Basin's NW to SE orientation, thrust bounded structural setting and foreland basin facies architecture are similar to many other ARM basins, suggesting any model for the ARM tectonic event should provide a mechanism for localized NE to SW contraction and the development of intra-foreland flexural basins.

The Oquirrh Formation

The Oquirrh Formation is an unusually thick miogeoclinal sequence of calcareous and clastic sedimentary rocks ranging from Morrowan (Early Pennsylvanian) through Wolfcampian (Early Permian age) in central and northern Utah and southern Idaho. The Oquirrh Formation lies conformably on the Manning Canyon formation of late Mississippian and early Pennsylvanian age, except in the Stansbury Mountains where the contact is an angular unconformity. The Oquirrh Formation is conformably overlain by Wolfcampian age limestones of the Kirkman Formation (**Figure 42**).

The Weber Formation is essentially the stable shelf equivalent of the Oquirrh Formation in northeastern Utah and northwestern Colorado. It is largely an unfossiliferous series of sandstone and orthoquartzite with a few interbedded limestone units ranging in age from Desmoinesian (late Middle Pennsylvanian) to Wolfcampian. It is 2260 feet thick at its type locality near Morgan in Weber Canyon (**Figure 43**).

The Weber thins gradually eastward to about 1000 feet at the Colorado border, eventually inter-fingering with the Maroon Formation to the east. The Weber disconformably overlies Pennsylvanian limestones and shales of the Morgan Formation and is disconformably overlain by the Permian Park City-Phosphoria Formations (**Figure 44**).

Oquirrh Basin South Mountain Section

The entire exposure of Oquirrh Formation at South Mountain west of Stockton in Tooele County, Utah was measured, described, and sampled. South Mountain contains a broad northwest plunging anticline with overturned strata on the nose of the fold.

Approximately 15,540 feet of Oquirrh strata is exposed here ranging from Morrowan through Wolfcampian ages based on fusulinid studies. The Wolfcampian portion of the Oquirrh Formation is 9410 feet thick (**Figure 45**).

The underlying Manning Canyon Shale is not exposed but the contact with the overlying Wolfcampian Kirkman Formation is at the base of the west slope of the western summit of South Mountain. The measured section is composed of a relative proportion of rock types consisting of gray to dark gray clastic limestones composed of detrital calcite grains and skeletal material in a matrix of micrite or fine calcitic mud. This calcarenite ranges from 73 to 98% carbonate, and the insoluble residue contains silt, clay, silica, and various skeletal fragments.

Figure 42. The Stansbury Mountains exposes the Manning Canyon Formation in angular contact with the overlying Oquirrh Formation. The foreground hills represent the Manning Canyon formation which dip to the west into the page while the overlying Oquirrh Formation along the ridgeline are relatively horizontally bedded.

Figure 43. Weber Canyon exposes the Weber Formation, a stable shelf equivalent to the Oquirrh Formation near Morgan, Utah.

Figure 44. The Phosphoria Formation in Wyoming is also known as the Park City Formation in Utah contains deposits of phosphate (black beds) in the above photograph. Source: Alchetron posted on the internet.

Figure 45. South Mountain, west of Stockton Utah represents the contact with the buried Manning Canyon Formation at the base of the mountains with the overlying Kirkman Formation at the summit of the western peak.

Cherty limestones are medium gray with light gray weathering, aphanitic to slightly silty, siliceous limestones containing nodules and stringers of massive concretionary bedded gray to brown chert in discontinuous and irregular distribution. The largest chert nodule observed was measured slighly more than eight feet across. Cherty limestones conisist of 7 to 88% carbonate with residue of angular fragments of dense, vitreous, dary gray to black chert. Gray, calcareous and platy shales weather to light brown colors and commonly split into quarter inch laminae. Gray, highly argillaceous platy limestones are gradational with the shale.

Light gray to light brown, friable non calcareous rocks composed of very fine quartz sand and silt are partially cemented with silica of calcite forming quartz sandstone. A typical sandstone consists of 97% quartz, 1% clay, and 1% cement with traces of microcline, detrital chalcedony, and heavy minerals. The sand ranges from very fine to silt size, and grains are well sorted, sub-rounded to well rounded, anhedral and equant. Grain boundaries are rough and quartz grains are occasionally intergrown.

An increased silica cement resulting in increased induration with tendencies to break around rather than across grains results in silicified sandstone. They are gradational between friable quartz sandstone and ortho-quartzite. Weathered surfaces are generally smoother than fresh rock. This rock is similar in composition to quartz sandstone except with a larger number of inter-grown quartz grains and a decrease in porosity to five percent or less. Original grain shape is increasingly obscured by secondary quartz.

Calcareous quartz sandstones have a significant amount of calcite present either as matrix or cement. Calcite is present as fine granular silt rather than as a chemical precipitate, filling interstices between quartz grains. A small amount of silica cement is also present. In the thin sections, bedding is commonly visible with layers of fine and very fine grained, well sorted, well rounded, equant anhedral quartz sand. Grain boundaries are highly etched where quartz is in contact with calcite, quartz inter-growths are rare. Calcite content is 20%. Insoluble residues consists of frosted grains of very fine quartz sand and silt occurring as isolated grains and as fragments of porous, friable sandstone and siltstone. Minute vein-lets of sparry calcite are also common.

Calcareous quartz siltstone consists of quartz grains which are silt size. Quartz grains are well sorted, equant, and angular to sub-angular in shape. The rock consists up to five percent second cycle quartz and quartzite or sandstone rock fragments. Carbonate content ranges from 10 to 37%, and the residue consists of light gray, vitreous silty ortho-quartzite, siltstone, and gray clay.

Light gray to light brown, non-calcareous silica cemented sandstone breaks with a smooth, sub-conchoidal fracture. Sand grains are not visibly present in hand sample and are fine to very fine grained quartz 95 to 100%, chert, chalcedony, orthoclase, or microcline. These rocks are typically homogenous, tenacious, and nonporous. Grains are well sorted, well bedded, and tightly inter-grown. Grain boundaries are smooth, curved, and deeply embayed. Carbonate content was three percent or less. These rocks fit into the ortho-quartzite class.

Calcareous ortho-quartzites with 3 to 30% interstitial calcite matrix are present with silica cement predominating while displaying intra-granular fracture is some units. Minor amounts of detrital microcline, orthoclase, chalcedony, and sandstone rock fragments are present. In many cases, silica cement and quartz intergrowths developed around detrital calcite grains providing ragged edges to the quartz. Insoluble residues include clear and frosted grains of quartz sand and silt, and fragments of porous quartz sandstone and siltstone.

Calc-arenaceous ortho-quartzites contain detrital calcite sand grains. The cement is silica with intra-granular fractures. Maximum calcite cement was 25% in a single unit. Calc-arenacous ortho-quartzites cannot be distinguished in the field from calcareous ortho-quartzites and both lithologies are gradational. In addition to quartz and calcite sand, a certain amount of calcite silt is present. Sand is very fine grained, sub-rounded to well rounded, and the quartz grains are interlocking with crenulated contacts. Insoluble residues consist of light gray to white, hard, dense, vitreous quartzite fragments and very fine quartz grains.

Quartzite breccia is composed of a wide variety of cata-clastite resulting from normal faulting. This rock is light brown, often friable breccia of very fine grained ortho-quartzite fragments, generally less than one centimeter in diameter. The breccia is scattered in a matrix of coarse, angular, quartz sand and silt. In rare cases, post brecciation re-cementation by silica occurred. Calcite content ranges from zero to ten percent in fresh samples. Weathered temples are generally leached of calcite. Insoluble residues of angular fragments of quartzite and quartz siltstone, clay, and silt are present.

Exposures are fair to good at South Mountain concealed by soils, landslide debris, and Tertiary lava flows representing six percent of the section. Minor variations in cementation as well as geomorphic vicissitudes can be responsible for the non-exposure of any interval in a stratigraphic section.

Hobble Creek Canyon Section

Wolfcampian strata of the Oquirrh Formation in Hobble Creek Canyon in Utah County were examined and sampled at regular intervals. Wolfcampian beds include nearly 7250 feet of ortho-quartzite, calcareous siltstone, and sandy limestone. These sequences bear striking resemblance to equivalent beds of South Mountain. Best exposures are on the south side of the south fork of Hobble Creek Canyon (**Figure 46**).

Lithologies

Ortho-quartzites are very fine grained, sub-rounded, interlocking grains of quartz up to 95% or greater, chalcedony, and microcline cemented by quartz and chalcedony. Rocks are ridge formers, hard, indurated, and split into smooth sub-conchoidal fracture. Although this type is non-calcareous in field examination, acid solution reveals up to four percent carbonate is present.

Calcareous and calc-arenaceous ortho-quartzite rocks consist of very fine grained silica cemented, quartz sandstone and siltstone containing up to 25% of fine calcite silt as matrix. These rocks break with a general intra-granular fracture, but the broken surface is not as smooth as in the non-calcareous ortho-quartzites. In thin section, more than half the quartz grains appear inter-grown, and grain boundaries are generally ragged and etched. A few random grains of well rounded, coarse quartz and calcite sand are present. Clay content is less than one percent.

Quartz sandstones are composed of predominantly fine to medium grained, light gray to light brown fairly to poorly sorted, friable, quartz sandstone with up to 30% porosity. This type is not common in the section and may represent weathered calcareous ortho-quartzite. Carbonate content is two percent or less.

Calcareous siltstones are light gray, weather brown, non-porous siltstone. Both weathered and unweathered portions are highly calcareous with up to 34% carbonate. The rock type resembles calcareous ortho-quartzite of which it is a gradational phase. Cement is mainly calcite with minor amounts of silica and up to three percent clay.

Sandy limestones are light gray, sandy to silty, light brown when weathered limestone including all gradations from sandy micritic limestone to bio-clastic calc-arenite forms. Occasional traces of authigenic opal and silicified bryozoan fragments, crinoid columnals, and miscellaneous skeletal material and siliceous casts of skeletal material are present.

Figure 46. Hobble Creek Canyon exposes the Oquirrh Formation on the south side of the canyon along the south fork of Hobble Creek on the north side of the creek.

Exposures of Wolfcampian age strata in Hobble Creek Canyon are generally good with approximately three percent of the section covered.

Weber Creek Canyon Section

The Wolfcampian interval of the Weber formation on the northeast margin of the Oquirrh Basin was part of the Weber Shelf during much of the Pennsylvanian and Permian time. Stratigraphic variation occurred from the central to the marginal part of the basin. Thickness of Wolfcampian strata at this location is 714 feet. The Weber Canyon section consists of relative proportions of rock types present in the section.

Principle lithologic changes extending from basin to shelf areas are absent of calcite in the matrices and cement of clastic rocks with lack of limestones in the section.

Lithologies

Ortho-quartzites are the most common type of rock in the Weber section. They typically consist of fine to very fine sand and silt, sub-angular to sub-rounded, anhedral quartz grains and silica cement. Grains have secondary quartz overgrowths and form an interlocking meshwork in which the interstices were filled with precipitated quartz. Porosity varies from a trace to one percent and is widely scattered clay line cavities that are one quarter millimeter or less in diameter. Grain boundaries are generally smooth and curved to slightly crenulated and embayed, and extinction is normal to moderately undulatory. Traces of detrital orthoclase, microcline, and chalcedony are present. These rocks have a sharp intra-granular fracture difficult or impossible to separate from meta-quartzite when comparing hand samples (**Figure 47**).

Figure 47. Weber Sandstone ortho-quartzite is exposed in these tilted beds uplifted during the Laramide orogeny. Source: Flickr posted on the internet.

Siliceous quartz sandstone is composed of rocks similar to the ortho-quartzites with less complete silica cementation and more rough, uneven inter-granular fracture.

Sandstone breccias are light gray to light brown breccias made up of fragments 3 inches long consisting of medium grained sandstone and ortho-quartzite in a matrix of friable, coarse, angular quartz sand. These occur as a result of local faulting of rocks, apparently have a low limit of plastic deformation based on the presence of ortho-quartzites accompanied by cata-clastites.

Exposures in Weber Canyon are good due to vertical strata and steep canyon walls. Covered intervals are about 16% of the section and occur mainly in zones of faulting and fault breccia.

Lateral Variations with the Oquirrh Formation

There are rapid facies changes in the Oquirrh Formation particularly in the Wolfcampian segment for individual strata that usually dies out or merges with adjacent strata within short distances. They seldomly persist for about ½ mile. Abrupt local changes in the character and degree of cementation were noted near the top of South Mountain. Bedding varies from apparently massive to thin and platy with quarter inch laminae or to other types of stratification within a few yards on a local basis. Such changes are believed to be due to subtle differences in cementation or amount and composition of matrices. The matrix consists of largely carbonate silt of detrital origin. The overall aspect of the section is generally unchanged from one canyon or exposure to the next at South Mountain. Individual lithologic units do not persist laterally. No definite patterns of lateral zonation were observed in the Oquirrh strata. Directions of facies changes appeared random.

The lensing character of Oquirrh strata is believed to be a depositional feature, not the result of post diagenetic erosion or strike slip faulting. Sediment may have been eroded by currents and waves prior to lithification, and re-deposited unevenly due to intermittent and irregular subsidence within the basin.

Quartz Types

Quartz grains in the Oquirrh Formation were used to determine sedimentation history. Common occurrence of detrital chalcedony grains, general deficiency of heavy minerals, presence of quartzite and sandstone fragments, scarcity of inclusions in quartz grains, and lack of euhedral crystal facies on quartz grains all indicate that these quartz sands are at least second cycle sediments derived from pre-existing sedimentary rocks.

Quartz grains in the Oquirrh Formation in the Stansbury Range west of South Mountain consist of angular to sub-angular silt size to fine grained quartz with normal extinction, and sub-rounded to rounded medium to coarse grained quartz with undulatory extinction. Both of these types were noted during the present study in varying amounts along with several other types. The grains are mainly silt size to fine grained sand with rough embayed or crenulated borders, moderate undulatory extinction with original grain shape hidden by secondary quartz.

The most common type of quartz was of sedimentary origin or was derived from sedimentary rocks based on the presence of rounded abraded secondary quartz overgrowths, carbonate inclusions, and dust inclusions outlining the shape of the original quartz grains. Plutonic igneous quartz, vein quartz, and pressure type metamorphic quartz were also present.

Weathering Properties

Surface alteration of calcareous and calc-arenaceous ortho-quartzites in the temperate, semi-arid climate of the eastern Great Basin commonly produces a concentric series of zones or rinds which are readily seen on broken rock surfaces. The outer zone is generally brown or dark brown porous, friable quartz sandstone from which calcite has been leached with clay and iron oxide content increased to about five percent. Particularly those rocks with lower original calcite content at South Mountain, the outer surface of this zone is case hardened with a thin veneer of smooth, dense silica. Contact of the weathered and un-weathered zones is fairly sharp and represents the maximum depth of effective water penetration. Weathered zones range from one eighth to several inches thick, but are usually less than two inches deep. Nature of the weathered surface is a function of lithology, climate, drainage, and length of time that the rock has been exposed. Zonation is strongest in the Hobble Creek Canyon likely due to increased precipitation and heavier vegetation at this locality but may also reflect subtle lithologic differences.

Porosity

Porosity of weathered zones may be as much as twenty percent or more, sufficient to permit migration of fluids where encountered in the subsurface.

Cementation

Cement found in sandstone is silica and carbonate. Sufficient concentration of pure quartz sandstones by silica under normal sedimentary conditions produces an ortho-quartzite. Cementation occurs by addition of silica to the environment, by pressure solution of quartz grains, or be a combination of these processes.

Addition of silica to sand under conditions of low pressure gives rise to euhedral overgrowths on the quartz grains in optical continuity with the original grain. Euhedral faces are likely to develop if the sand is porous and loosely packed, and if the grains do not interpenetrate each other. Silica may originate as a product of mineral alteration such as alteration of montmorillanite to illite in interbedded clays, or by replacement of quartz or chert by carbonate, or from connate water or ground water.

Pressure solution phenomenon develop increased pressure on a solid reduces the melting point of that solid for any given set of chemical conditions. Quartz dissolves at the points of highest pressure especially in zones of high pH and is precipitated at points of lower pressure and low pH. Presence of interstitial clays and water is important. The clay contributes potassium ions which are replaced by calcium and magnesium ions through base exchange, and by potassium carbonate forms. This raises the pH in clay rich zones. The water aids silica solution, acting as a vehicle or migration of the dissolved silica. Compaction then proceeds and grains become tightly packed and are usually interpenetrating.

Cementation of the ortho-quartzites in the Oquirrh Formation at South Mountain and at Hobble Creek Canyon occurred as a result of combined pressure solution and addition of new silica. The majority of ortho-quartzite thin sections examined shows quartz grains welded together both by interpenetration and by deposition of secondary silica in pore spaces as rims or overgrowth on original quartz grains. Grain contacts are in the absence of carbonate, sinuous, curved, and sutured. In calcareous and calc-arenaceous ortho-quartzites, the effect of calcite has been development of rough, etched grain borders. Inter-grown quartz grains still predominate.

The Weber Formation thin sections also show both pressure solution and secondary overgrowths although euhedral overgrowths are rare. Grain boundaries are embayed and crenulated to the extent that original grain shapes are obscured.

Implications are that there was a plethora of silica in the Oquirrh Basin during deposition of the Oquirrh Formation or shortly after, and that strong tectonic forces were active in the area. The abundance of silica makes it unlikely that zones of good porosity would be encountered in Wolfcampian strata of the formation in the subsurface.

Tectonic Setting

The Oquirrh Basin was part of the Cordilleran Geosyncline during Early Paleozoic time. From the Devonian until the end of the Paleozoic, the older orthogeosyncline was divided by a geanticline which arose in western and central Nevada, effectively separating the western eugeosyncline from the eastern miogeocline. The western limit of the late Paleozoic miogeocline is called the Manhattan Line, bisecting the state of Nevada into a nearly north to south direction along the Antler-Sonoma orogenic belt. The eastern limit was a line of flexure called the Wasatch Line which follows the eastern edge of the Great Basin geomorphic province. The Wastach Lline separated the miogeocline from the stable shelf and cratonic areas to the east. This hinge line is a narrow zone of rapid change in sedimentary thickness and lithology, and it is across this line that the Oquirrh and Weber Formations interfinger. It is immediately west of the Wasatch Lline in the eastern part of the miogeocline that the thickest known section of Pennsylvanian and Permian rocks are found anywhere in the United States with the possible exception of the Ouachita Mountain region of Oklahoma and Arkansas.

End of the miogeocline came towards the close of the Permian Period. Late Permian rocks are unknown in the area and Triassic strata were deposited unconformably on Permian rocks.

Depositional Environment

The overall depositional pattern in the Oquirrh Basin was one of clastic sedimentation including both quartz and calcite sand, and silt with intermittent shorter periods of limestone deposition. Lithologic variability, abrupt changes in thickness of individual units, and abundance of clastic materials indicate that Pennsylvanian and Early Permian was a time of general in-stability. Subsidence was continuous but uneven with considerable local adjustment. It is possible that littoral and lagoonal sediments were deposited at various places within the basin where much of the sequence was destroyed by erosion. That erosion occurred at various times is suggested by numerous small unconformities and diastemic breaks observed at South Mountain and in Hobble Creek Canyon.

Environments ranged from intra-neritic to epi-neritic and epi-neritic bio-stromal with water depth variable but probably fewer than a few hundred feet. Sediments are fairly well sorted and rounded, coupled with deficiency in clay size material. This indicates long periods of washing and transportation over long distances. Current ripple marks and small scale cross bedding are apparent on weathered surfaces. Silica cementation tends to obscure these and other primary structures. Areas of thickest accumulation of sediment varied from epoch to epoch.

Precipitation of silica cement in an environment where carbonate sediments are forming requires precise geochemical conditions. Silica does not precipitate above a pH of 8.8. Calcite does not precipitate below a pH of 8.0. Oxidation potential may be either positive or negative and does not appear to be a controlling factor.

Paleogeography

Borderland areas adjacent to the Oquirrh Basin during the Pennsylvanian and Early Permian include the Weber Shelf to the northeast, the Northeast Nevada Highland to the northwest, the Western Utah Highland to the west and southwest, the Callville-Hermosa Platform to the south, and the Emery Uplift to the southeast. These positive areas were breached at various places by inlets or access ways which permitted contact with open ocean waters through which considerable amounts of sediment may have been transported to the basin.

During Middle and Late Wolfcampian time, there was a pronounced decrease in size of the Oquirrh Basin probably due to an interruption of the relatively rapid rate of subsidence experienced earlier. Eolian sediments of the Diamond Creek and upper Weber Formation covered the Weber Shelf and the littoral and epi-neritic Kirkman Limestone which was deposited over the Oquirrh Mountains.

Paleontologic Evidence

Limestone units of the Oquirrh Formation contain a varied fauna principally consisting of fusulinids, crinoids, bryozoans, brachiopods, corals, and various forms of algae. The most widely distributed are the fusulinids. Fusulinids are predominantly benthonic calcareous foraminifera which apparently were tolerant of a wide range of environmental conditions. Rock types in which they are found include a wide variety of limestones, sandy limestones, and calcareous shale and siltstone. Normally, fusulinids thrived in fairly quiet marine waters at depths considered to range from one to thirty fathoms. Variation in fusulinid litho-topes, and their common occurrence in layers barren of other organisms indicates they were tolerant of greater ecological extremes.

Occasional local abundance of corals, crinoids, algae, brachiopods, and bryozoans indicates at times the water was shallow enough to permit penetration of sunlight, sufficiently agitated to provide oxygen and nutrients. Waters were also clear enough to permit these forms to flourish. Shifting environments oscillating between littoral, epi-neritic, and infra-neritic conditions with moderate to strong bottom currents, and abundant clastic supply with continuous subsidence is compatible with paleontologic evidence.

Rate of Sedimentation

The Pennsylvanian Period lasted 55 million years. The average rate of Oquirrh sedimentation is 290 feet per million years, or 3433 years per foot of sedimentary rock. This rate is greater than the maximum miogeoclinal rate but less than the maximum rate for intra-cratonal geosynclines. The Oquirrh Basin was more active than most miogeoclinal basins, and more similar to intra-cratonal basins in its tectonic behavior.

Comparisons with Quartzites from Other Formations

The variety of calc-arenaceous ortho-quartzite predominating in the Oquirrh Formation has not been found elsewhere. A review of sedimentary petrology texts along with periodical literature failed to produce similar lithologic rocks. Approximately 35 formations from different areas and several systems were reviewed. Two dozen thin section were prepared from these formations but failed to reveal any other examples of calc-arenaceous ortho-quartzite. Similarities in texture, nature of the quartz grains, and type of cement were noted but the combination of fine and very fine quartz sand, carbonate silt, and silica cement was not found.

Quartzite Observations

Thin section studies from the Oquirrh, Weber, and other formations indicate characteristics of quartzite lithology. Petrographic properties are described in the following discussion.

The nature of cement in ortho-quartzites are mainly a combination of pressure solution of detrital quartz grains and addition of new silica with occasional examples where these processes occurred separately.

The detrital fraction of high quartz content ranges from more than 95% of the thin sections analyzed which contained at least 95% quartz including detrital chert grains and quartzose rock fragments. More than 91% contained 99% quartz.

Light color occurred in more than 75% of samples, and were mega-scopically consisting of light shades of gray, brown, pink, with none of the thin sections measurably colored. There was a complete lack of effective porosity in ortho-quartzites. More than 90% of samples studied had less than 1% isolated pore spaces and no pore spaces occurred in 5% of the samples.

Idio-morphism in detrital quartz consisted of all anhedral grains. Occasionally, grains of subhedral vein quartz were found. In some cases, crystal faces may develop on secondary quartz which formed in veins or cavities in an ortho-quartzite after diagenesis.

Some properties were dissimilar in ortho-quartzites which were consistently variable such as grain size, sorting, roundness, and nature of grain boundaries, inter-granular, and inclusions occurred in the quartz grains.

Ortho-quartzite vs. Meta-quartzite

Differentiation of ortho-quartzite from meta-quartzite was traditionally dependent on a study of field relationships of the rock involved or the identification of characteristic metamorphic mineral which may be present in any given sample. Ortho-quartzites are generally non-foliated, bedding and cross beddings may or may not be present; detrital grain outlines may be visibly enclosed in secondary silica with a possible degree of rounding; grains may be tightly packed and inter-grown or loosely packed with a fair amount of lack of cement porosity; most quartz grains show moderate strain shadows with some having normal extinction; fractures are commonly observed in thin section cutting across grains and cement as deformation structures; and faunal evidence may appear as unaltered plant or animal fossils.

On the other hand, metaquartzites show a parallel to sub-parallel orientation of new minerals such as micas, provided the material is availabe for development of these minerals as foliation; original grain shape is obscured by recrystallization and the absence of dust rims; grains are tightly packed, intergrown and interlocking with crenulated or sutured grain contacts; extinction is highly undulatory with strong strain shadows; shear may be present in crushed, strung out, and recrystallizaed grains showing deformation; and preservation of fossils is highly unlikely.

Since the process of metamorphism is gradational with those of diagenesis, no sharp line is drawn between orthoquartzite and metaquartzite. Some criterai described for metaquartzite may develop under diageneisis to some degree but may also occur in orthoquartzite.

Summary

Calcarenaceous orthoquartzite in the Oquirrh Formation is a lithologic unique stratigraphy. Calcareous sediment in the calcarenaceous orthoquartzite is probably derived from the various highlands adjacent to the Oquirrh Basin and contributed from outside the basin. Much of this is second cycle sediment. A source for the chemically precipitated silica is the volcanic area in the Cordilleran eugeocline to the west. Conditions prevailed within the basin permitted chemical precipitation of both calcite and silica. These conditions persisted throughout most of Pennsylvanian and Wolfcampian time while sufficient subsidence was occurring so that nearly five miles of sediment accumulated.

Conclusions

The Oquirrh Formation constitutes an unusual suite of sedimentary rocks which reflect conditions of tectonics and sediment supply perhaps unique in geologic history. The Oquirrh Formation is unique in its great thickness, in the lack of shale or claystone, and in the abundance of calcarenaceous orthoquartzite.

Shallow water marine conditions prevailed in the Oquirrh Basin during Pennsylvanian and Early Permian time in conjunction with tectonic conditions of irregular instability and deep subsidence.

Sediment of Oquirrh Formation was derived from various adjacent land masses, from local positive areas within the Oquirrh Basin and perhaps also from beyond the basin through peripheral access ways. This material was subjected to effective sorting and cleaning by wave and current action.

Orthoquartzite and calcarenaceous orthoquartzite of the Oquirrh Formation are rock types which do not fit neatly into the present system of sedimentary rock classification. Caution should be exercised when describing these rocks or when studying published descriptions of Oquirrh strata.

References

Baars, D.L., Stevenson, G.M. 1981. Tectonic evolution of western Colorado and eastern Utah in Western Slope (Western Colorado), Epis, R. C.; Callender, J. F.; [eds.], New Mexico Geological Society 32nd Annual Fall Field Conference Guidebook, 337 p.

Casillas, H.A. 2004. An integrated geophysical study of the Uncompahgre Uplift, Colorado and Utah. Department of Geological Sciences, University of Texas, el Paso.

Nuccio, V.F., Condon, S.M. 1996. Burial and Thermal History of the Paradox Basin, Utah and Colorado, and Petroleum Potential of the Middle Pennsylvanian Paradox Formation. Evolution of Sedimentary Basin – Paradox Basin. US Geological Survey Bulletin 2000-0.

Soreghan, G.S., Keller, G.R., Gilbert, M.C., Chase, C.G., Sweet, D.E. 2012. Load-induced subsidence of the Ancestral Rocky Mountains recorded by preservation of Permian landscapes. Geosphere; June 2012; v. 8; no. 3; p. 654–668.

Wells, R.B. 1963. Orthoquartzites of the Oquirrh Formation. Geology Studies, Vol. 10. Brigham Young University, Department of Geology, Provo, Utah.

www.ingramcontent.com/pod-product-compliance
Ingram Content Group UK Ltd.
Pitfield, Milton Keynes, MK11 3LW, UK
UKHW060400300726
14090UKWH00001B/43